THE KEW GARDENER'S GUIDE TO

GROWING
ALPINES

THE KEW GARDENER'S GUIDE TO

GROWING ALPINES

THE ART AND SCIENCE TO
GROW WITH CONFIDENCE

MATTHEW JEFFERY

FRANCES
LINCOLN

Contents

Introduction to growing alpines

—

THE VALUE OF ALPINES

The term 'alpine' covers an incredibly diverse range of plants from right across the globe. Because of their generally small size, you can fit plants from all over the world into a confined space. They are perfect for pots, balconies and even outdoor windowsills. In a world increasingly affected by a changing climate, carefully selected alpine plants can provide solutions for drought and sun. They are often very easy to care for, not requiring much pruning, watering or feeding, and they create very little in the way of leaf-litter mess. Thus, alpines are ideal for modern busy lifestyles.

You may have already been introduced to an alpine without realizing. The first alpines I saw were houseleeks (*Sempervivum*) that my grandfather grew in small pots – a plant that I later discovered was from the mountains of Europe.

WHAT IS AN ALPINE?

A true alpine is defined as a plant growing above the treeline in the wild; this can be at high altitude, and also high latitude. High-altitude environments are characterized by having raised levels of ultraviolet (UV) light, and increased exposure and wind due to lack of trees. They are generally cooler than lower altitudes, but they also experience extreme temperature fluctuations throughout the day and the year, often with deep snow in winter, shortening the growing season. High-latitude environments (nearer to the poles) are subject to similar conditions but, instead of constant high UV, they have a short and intense growing season with much of their yearly daylight condensed into a few months.

The terms 'horticultural alpine' and 'rock-garden plant' are more appropriate for the wider range of plants that we are accustomed to grow in areas of cultivation. They can include plants from all over the world and different environments, but, generally, all are small in habit and are hardy. Rock-garden plants are not all from alpine environments, so they can be incredibly useful in the garden for further extending the flowering season and providing plants with longer individual

The Mediterranean sea holly (*Eryngium bourgatii*) thrives in the high-altitude meadows of the Picos de Europa mountain range in northern Spain.

This beautiful alpine meadow at 1,716m/5,630ft in the French Alps is dominated by wild buttercups (*Ranunculus* spp.) and cranesbill (*Geranium* spp.).

flowering. This book covers the range of plants that you can grow in rock gardens, pots and troughs, as well as those you are likely to find in the 'alpine' section of the garden centre (even if some are from a coastal sea cliff).

WHERE TO SEE ALPINES

Alpine plants can be found all over the world: the alpine ecosystem includes habitats such as arctic tundra, rocky mountain peaks, high-altitude meadows, scree (slopes of loose rocks) and steep grassy hillsides. Many of the places alpine plants grow are easily accessible; many are in national parks.

There are rock gardens spread across the world, including in the UK, North America and Europe, while alpines are also likely to be in cultivation much closer to where you live, often in your local botanic garden. Alpine plant societies, such as the Alpine Garden Society, the Scottish Rock Garden Club and the North American Rock Garden Society, often host talks, plant shows and events based on the cultivation of alpine plants.

Alpine conservation

Alpine ecosystems are increasingly under threat from a wide range of factors. Tourism can be a double-edged sword: it can bring much-needed attention to beautiful areas in need of protection, but it can also increase footfall and demand for the land. Many upland areas are particularly popular for winter sports and summer mountain activities such as hiking and biking. These have been instrumental in increasing accessibility to otherwise inaccessible areas, enhancing appreciation and awareness. When visiting mountain environments, it is especially important to remain on existing paths/roads/ski pistes, not only for personal safety but also to minimize ecological damage.

Another main factor that is negatively impacting the health of these alpine ecosystems is climate change. With global temperatures rising, these cool-adapted plants are being outcompeted by increased growth of other plants moving into their ecological niches and forcing them to higher and higher altitudes. Soon there will be nowhere for them to go. Although this process is now irreversible without a large global shift in emissions, there are other risk factors that can be influenced on a more local scale. One of the next largest threats is increased grazing pressure from wild animals and livestock. Without proper management, wild deer and domestic livestock can reduce wild plants so they exist on only the most unreachable rock faces.

Heavier industrial activities such as mining, quarrying and the construction of buildings, ski lifts and even wind farms can cause immediate damage and long-term environmental consequences.

There are many conservation activities, both at the local and international scale, that are working to safeguard the flora of alpine ecosystems. The Royal Botanic Gardens, Kew, is involved in many such initiatives for research, conservation, *in-situ* and *ex-situ* conservation. Partnerships exist across the globe, collecting data, seeds, plants and DNA for conservation and research on alpine plants. Kew's Millennium

Staff from the Royal Botanic Gardens, Kew, together with staff from Graz Botanical Garden, collect seed in the Austrian Alps for the Millennium Seed Bank Partnership.

Seed Bank, at Wakehurst, contains tens of thousands of seeds from alpine plants across the world; these are actively used for research into understanding alpine plants, and are also a global seed resource and safeguard against species extinction.

Ongoing alpine research at Kew and with global partners includes topics as diverse as Red List assessment (such as the shifting ranges of species and risks to species), fungal root (mycorrhizae) relationships, seed-coat microbiomes, genetics of alpine plant populations and the adaptation of alpine plants.

WAYS THAT ALPINES HAVE ADAPTED

Many true alpines have evolved specific evolutionary adaptations to survive the barrage of difficult growing conditions. A considerable number of adaptations and plant species are shared between high-altitude and high-latitude environments across the planet.

In the short alpine growing season, plants must make a choice. They can either hunker down while they try to survive, investing heavily in structures that can withstand the elements; some of these are underground. Or else they can live fast and die young, by having a one- or two-year life cycle (annual or biennial) and by producing lots of seeds.

Regardless of the strategy that each alpine species has evolved, its general size and shape are often very low to the ground with a compact habit. It will also have long, deep roots to anchor it and find water. Such features prevent alpines becoming damaged or uprooted in high winds and from rock fall in the ever-changing mountain environment.

Chiltern gentian (*Gentianella germanica*) thrives in the Austrian Alps at 2,000m/6,600ft above sea level. Its low-growing stature makes it wind-resistant, and its clusters of bright flowers attract pollinators from far and wide.

Mountains are all or nothing when it comes to water. At the start of the season, there will be lots of snowmelt to provide water, but this gets increasingly less towards summer. Most alpines, therefore, have adaptations to reduce water loss in their foliage: many develop leaves that are small, thin, succulent or hairy – and often in the form of a rosette where the foliage is tightly packed on the stem, forming a spiral. Many leaves are also spiny, to reduce predation from hungry alpine herbivores.

The main exception to their generally conservative growth is when alpines need to reproduce, and so they bear some of the most beautiful flowers in the plant kingdom. They must compete for pollinator attention using bright colours, unusual shapes and wonderful fragrance. Blooms are often disproportionately large when compared to the plant's size, and many are held at ground level, so they aren't broken by the wind. Conversely, some species risk the wind and grow their flowers on tall stalks to attract pollinators from afar. Because of the short growing season in alpine and arctic habitats, many true alpine species have individually short flowering seasons. However, in a diverse range of habitats across the world, when one alpine finishes blooming another takes over, right through the growing season.

This whole suite of adaptations is shared not only with many unrelated plants across alpine habitats but also with those that grow on coastlines, cliffs and deserts. A parallel range of environmental pressures has resulted in similar adaptations arising in unrelated plants: this is called convergent evolution.

Alpine adaptations

Hairy leaves
(save water; wind/UV resistance)

Spanish hawkweed (*Hieracium bombycinum*)

Low-growing habit
(provides wind resistance and temperature regulation)

Showy/scented flowers
(attract scarce pollinators)

Small leaves
(save water; wind/UV resistance)

Mountain avens (*Dryas octopetala*)

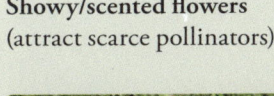

Garland flower (*Daphne cneorum*) and Horseshoe vetch (*Hippocrepis comosa*)

Dolomites saxifrage (*Saxifraga squarrosa*)

Spines
(provide protection from herbivores)

Succulence/thick leaf cuticle
(saves water; wind/UV resistance)

Underground storage
(provides protection from the elements and saves water)

Pyrenean thistle (*Carduus carlinoides*)

Common houseleek (*Sempervivum tectorum*)

Ivy-leaved cyclamen (*Cyclamen hederifolium*)

Alpine life forms

Annual or biennial

Monocarpic biennial:
Yellow bellflower
(*Campanula thyrsoides*)

Climber

Alpine clematis
(*Clematis alpina*)

Cushion-forming plant

Moss campion (*Silene acaulis*)

Dwarf shrub or subshrub

Net-leaved willow
(*Salix reticulata*)

Evergreen perennial

Trumpet gentian
(*Gentiana acaulis*)

Geophyte

Bulb: Dog's tooth violet
(*Erythronium dens-canis*)

Succulent

Cobweb houseleek
(*Sempervivum arachnoideum*)

Herbaceous perennial

Snowbell (*Soldanella alpina*)

Mat-forming plant

Creeping globe daisy
(*Globularia repens*)

ALPINE LIFE FORMS EXPLAINED

Annuals and biennials

Annual plants live fastest and die the youngest; they condense their whole life cycle into the short growing season from seed to seed in just a few months.

Biennials differ by often making a strong plant in the first year, then overwintering and bearing flowers in the second year. Both these life forms often produce vast quantities of seeds that can survive in the ground for many years, ungerminated, to ensure against bad years.

Annuals and biennials are monocarpic, which means after flowering and setting seed the plant will generally die. Some monocarpic species live longer than two years before gaining enough energy to flower and produce seeds.

Climbers

Climbing is not a very common strategy for alpine plants as there is often not much to climb. However, some climbers still grow in mountainous regions on the borders of the alpine zone, where there are trees and shrubs. You can find climbers such as the alpine clematis *Clematis alpina* in the European Alps growing not only up trees and shrubs but also along the ground in between rocks.

Cushion-forming plants

The cushion life form is seen in alpine zones across the world in unrelated plants. It is a highly specialized and effective temperature-regulation strategy that retains heat in the centre of the plant when the weather is cold

and windy, but keeps it cool in the centre when conditions are hot and sunny. Many cushion-forming plants are also evergreen, but they retain all their old leaves inside the cushion to form a very solid plant. See also Propagating a cushion-forming plant, page 66.

Dwarf shrubs or subshrubs

The only way to survive the wind and snow as a shrub is to stay small or very flexible. Dwarf shrubs often stay very low to the ground and can be evergreen, such as Alpine juniper (*Juniperus communis* var. *saxatilis*), or deciduous, such as net-leaved willow (*Salix reticulata*).

Evergreen perennials

It is expensive to grow new leaves each year, especially when nutrients are scarce, so some species retain their leaves over winter; if they are small or tough enough, they aren't damaged by the winter weather. Such plants can be treated in a similar way to herbaceous perennials and be divided by the same method (see Dividing an herbaceous perennial alpine, page 54).

Geophytes: Alpine bulbs, tubers and corms

A geophyte is a plant that has its main growth organ sheltered safely underground as a tuber, bulb or corm. Some of our favourite true alpines are bulbs: for example, many species tulips (*Tulipa*), fritillaries (*Fritillaria*) and squill (*Scilla*). Some are corms, such as *Crocus*, and some are tubers, such as *Cyclamen*. These can be the first plants to flower in spring, and the last in autumn, making perfect companions to other alpine plants in the garden. The alpine collection at the

The potted spring display in the Davies Alpine House at Kew is full of colour from alliums, cranesbills (*Geranium*), *Leucocoryne* and saxifrages (*Saxifraga*).

Royal Botanic Gardens, Kew, contains many geophytic species from all over the world and is to be found spread between the potted nursery collections and the Rock Garden itself. See also Planting a patch of early squill bulbs, page 118.

Succulents
Retaining water in leaves is a great way to ensure a plant has a reserve to survive through windy and dry conditions. Such succulent plants have evolved from many different plant families not only in alpine ecosystems but also in deserts, cliffs and salty coastal habitats.

Herbaceous perennials
This is a very common life form for much of the temperate world, as we are accustomed to growing many herbaceous plants in general garden settings. Although an herbaceous plant is deciduous and dies back every year, its root system survives, often for many years. In this way such plants can avoid the more extreme winter weather. Some perennial alpine plants can be quite short-lived and survive for only three years or so, behaving almost as if biennial.

Mat-forming plants
These are essentially flattened and looser-growing versions of a cushion-forming plant (see page 15). A habit of growing as a mat is more common as a life form than a true cushion. Mat-forming plants can be evergreen, deciduous, woody or non-woody. Being close to the ground

Hardiness zones

The Royal Horticultural Society (RHS) has devised a simplified system for categorizing plant hardiness; this is summarized directly from the RHS in the table below. These are what I have used to estimate a plant's temperature tolerances, although this can vary depending on local microclimate, soil conditions and moisture.

RHS hardiness zone	Minimum temperature range	Category
H2	+1°C/34°F to 5°C/41°F	Tender – cool or frost-free greenhouse
H3	–5°C/23°F to +1°C/34°F	Half-hardy – unheated greenhouse/mild winter
H4	–10°C/14°F to – 5°C/23°F	Hardy – average winter
H5	–15°C/5°F to –10°C/14°F	Hardy – cold winter
H6	–20°C/–4°F to –15°/5°F	Hardy – very cold winter
H7	colder than –20°C/–4°F	Very hardy

Although many of the plants described in this book come from mountainous and rocky places, some may need protection from the worst wet or cold weather in climates they are not native to.

makes them very wind-, snow- and rock-resistant, as well as conferring some of the temperature-regulation capabilities of a true cushion-forming plant.

WHAT CAN YOU GROW, AND WHERE?

Local climatic conditions will play a large role in what you are able to cultivate outside. In areas of the world with warmer and drier climates, rock gardens may need to rely on a suite of more Mediterranean mountain plants; this would be in areas where summer temperatures often go above 30°C/86°F. In cooler and wetter climates where average temperature is above 30°C/86°F for only a short period, many true alpine plants will grow very well outside if drainage is adequate.

Rock gardens

Rock gardens are areas or gardens built primarily from stone, dedicated to alpine cultivation. The main purpose of using rockwork in the garden is to provide niches in which plants can thrive. You can create frost-free areas, pockets of shade,

Rockery vs rock garden: there is no hard-and-fast rule about what to call the space in which you grow alpine plants. Rockeries tend to be small, while rock gardens can be larger and more immersive (like this one at the Royal Botanic Gardens, Kew).

good drainage or moisture, all with rocks and growing media. Rock gardening, as we know it, really took off with Reginald Farrer in the late 1800s to early 1900s. Farrer took great influence from the rock landscapes of traditional Japanese gardening styles and from his visits to alpine ecosystems all over the world. Water features are a large part of what makes rock gardens feel so natural, with the sound and cooling nature of running water.

The Rock Garden at Kew was created in the 1880s using various rock types until being slowly changed to the current sandstone between 1929 and 1970. It is one of the largest and oldest rock gardens in the world.

Crevice gardens

Crevice gardens are a specific type of rock garden, in which flat pieces of stone are arranged vertically and closely set to mimic natural rock formations. This creates a buffer to the temperature and humidity fluctuations outside the crevice and encourages deep rooting alongside the cool, moist stone surfaces buried in the structure. Sand is often used as a growing

Crevice gardens are especially popular in Czechia, and the RHS was fortunate to engage the famous Czech crevice gardener Zdeněk Zvolánek to advise and help install the large crevice garden at RHS Garden Wisley, shown here. This is built to mimic natural landscapes.

media in crevice gardens as it doesn't break down and it allows water to pass through freely. For many people, a full-sized crevice garden may be unrealistic, but you can create a miniature one in a pot or trough (see Planting up an alpine crevice trough, page 48).

Raised beds

This is often the most achievable method of outdoor alpine cultivation in a garden setting. Many gardens already have raised beds, and it does not require any additional stone being imported. Raised

beds can be as little as 30cm/12in above the rest of the ground to be incredibly effective at increasing drainage. They can be made from wooden shuttering, railway sleepers, stones or bricks. They bring the plants closer to the gardener for care and enjoyment.

Pot cultivation

Growing alpines in pots is very traditional in alpine horticulture. Pots allow you to have a very tight control on the moisture level the roots experience – this is useful for tricky plants. Groups of pots also make

Terracotta pots filled with alpine succulents, including houseleek (*Sempervivum*) and *Rosularia*, set in a range of heights help make a very attractive display in the Davies Alpine House at Kew.

great seasonal displays, such as those in the Davies Alpine House at Kew. You can group plants together that require different watering regimes and get up close and personal with flowers. Pot plants are also easily transported to alpine shows and events.

The pot type of choice is often terracotta – this is not just for aesthetic reasons. The naturally porous nature of terracotta allows evaporation from the surface, constantly cooling plant roots. Terracotta also allows for effective use of a sand plunge (see Using a sand plunge for alpine plant cultivation, page 128). For species that require constant moisture, plastic pots may be a better choice.

A group of Candia tulips (*Tulipa saxatilis*) flowering in
the Rock Garden at Kew make a bold statement in front
of the Davies Alpine House.

At least three different species of lichens have covered this old stone trough after years of being left outside.

This artificial stone trough made from hypertufa contains *Sempervivum* 'Matthew's Day Dream' and *S.* 'Reinhard', as well as a small cutting of *Sedum sexangulare*.

Trough cultivation

Troughs are the ultimate way to make a miniature alpine habitat in a small space. As well as expensive real stone, they can also be made of artificial stone (see Casting a hypertufa trough, page 100) or from an old sink – even from a fish box covered in cement. In much the same way as pot cultivation (see page 19), troughs allow for more regulation of watering; just make sure they have drainage holes.

Weathering troughs

Lichens are a natural part of a rocky ecosystem. There are many ideas on how to grow lichens on troughs, but ultimately it is only a matter of waiting for them to appear naturally. Lichen colonization is faster on natural surfaces than artificial ones.

Winter protection, alpine houses and sand plunges

Some alpine species may require protection from winter wet, as they come from climates where winters are very dry and cold. Such plants are used to being covered under a deep layer of insulating (and dry) snow.

Winter protection can come in a variety of forms, and the main aim with alpines is to shelter them from wet, not from cold. This can be achieved by suspending a sheet of polycarbonate over an individual delicate alpine planted in the ground or in a trough that is too heavy to move, or by moving it into a cold frame, well-ventilated greenhouse or alpine house, before the offset of cold weather. If you live in an area that

The traditional alpine house (here, at the Royal Botanic Garden Edinburgh) has brick sides, side vents and sand-filled benches, where pots can be buried to keep them cool.

experiences heavy snow, a shelter for outdoor plants may need to be more substantial than a sheet of polycarbonate, to bear the additional weight.

Traditional alpine houses have benches filled with sand, known as sand plunges, in which plants grown in pots can be buried up to their rims if required. Especially when growing plants in terracotta pots, this can be advantageous for alpine plant growth. The sand can be moistened, and the water should wick through the porous terracotta to keep the sand slightly moist. This creates consistent cool humidity at the roots of the plants. See also Using a sand plunge for alpine plant cultivation, page 128.

The cushion-forming plant collection and the bulb collections (here of *Roscoea*) are grown in sand plunges behind the scenes in the alpine nursery at the Royal Botanic Gardens, Kew.

Rocks, rocks, rocks

The hard landscaping is really the most integral part of making a rock garden. Fortunately, the primary structure doesn't just have to be freshly quarried virgin stone. Increasingly, modern designs incorporate waste industrial stone or recycled concrete materials. If choosing real stone, try to find some to match your local geology. Whatever stone you select, it is important to obtain it as locally as possible, to reduce the environmental impact and cost of transportation – and use second-hand stone wherever possible. The different types of stone have their own advantages and disadvantages.

Sandstone is often the stone of choice, and it is used in the Rock Garden at RHS Garden Wisley and at the Royal Botanic Gardens, Kew. It is easy to cut and split. Limestone was once the preferred stone for rockeries, emulating the habitat with the most diverse alpine communities occurring on this slightly porous, alkaline rock.

Recycled concrete slabs and old slate roof tiles can be another good option, especially for crevice gardens (see page 18) as they already have a flat shape.

Granite has the advantage of being hardwearing, but the disadvantage of being very heavy.

The way rocks are arranged in a garden can take inspiration from rock formations in the wild. Here are two examples of natural rock formations: a sandstone outcrop and a limestone scree in the Austrian Alps.

Tufa is a beautiful natural stone formed relatively quickly (geologically speaking) around calcium-rich springs, lakes and streams. The roots of alpine plants can easily penetrate tufa to access water and minerals. Plants grown on it develop very densely because of a lack of nitrogen; thus, their habit is more akin to how they would look in the wild. Because of its porous nature, tufa has long been used as a surface on which to plant 'difficult' alpines. Its honeycomb structure enables it to draw up water like a sponge. The water constantly evaporates from the surface, so the tufa naturally cools itself in warm weather.

This stone is an important part of the ecosystem, and only small amounts can be taken at any time, to avoid ecological damage. It is, therefore, almost impossible to get hold of large enough quantities to build a whole rock garden. It is much better to buy second-hand tufa online; this may also have the advantage of including some established plants or pre-made planting holes. If you cannot get real tufa, you could make some hypertufa rock (see Casting a hypertufa trough, page 100). This will behave in a similar way to natural tufa, especially if you increase the proportion of compost in the mix to create a very porous artificial stone. (See also Planting into tufa, page 112.)

Planting in tufa can look very naturalistic as the plants grow directly from the stone itself. Here, a group of encrusted saxifrages (*Saxifraga* section Ligulatae) are growing from tufa in a trough at the Royal Botanic Gardens, Kew.

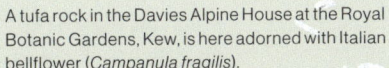

A tufa rock in the Davies Alpine House at the Royal Botanic Gardens, Kew, is here adorned with Italian bellflower (*Campanula fragilis*).

BUILDING A ROCK GARDEN

Having your own rock garden is a great idea for a small garden. It can be any size you like and will enable you to grow many plants that require better drainage for outdoor cultivation.

Aspect is an important consideration in the planning process: a cool aspect will allow you to grow shade-loving plants, while a sunny aspect will favour sun-lovers.

Once you have chosen your site, you need to create some height. This can be done with hardcore, bricks, stone, coarse grit or sand and should be at least 20cm/8cm high. On top of this layer, you can add a water-permeable, biodegradable membrane such as hessian, to prevent the soil from filling up the air gaps in the drainage layer. Cover the hessian with a 30cm/12in layer of your growing media mix (see Potting composts, soils and topdressings, page 29).

You will also want some rocks (see Rocks, rocks, rocks, page 24). It may help to draw a rough sketch of how you would like your rock garden to look; mark where you want the rock layers to go. The rocks are there to create niches and microclimates for the plants to sit between. There will be small areas of shade and cool crevices for roots to flourish, while keeping the neck and crown of each plant dry above the soil level. If you find that you need more rocks than you initially anticipated, it is better to adapt your plans, using rocks in high density, than to spread the rocks too thinly. You can also utilize existing features such as a brick wall or concrete edge to build against.

When working with heavy materials, it is important to consider personal safety. Thick gloves and steel-toe boots are ideal. If you cannot lift each stone comfortably, then seek professional contractor help.

Using a trowel, dig spaces for the rocks one at a time, starting with the largest ones at the bottom. Position each so it leans back into the mound, for stability, and bury it one-third to one-half into the growing mix; this encourages the plant roots to trace the edges of the rocks down into the ground. Backfill with the growing mix as you go, and firm in well. Use the trowel to ensure there are no large air gaps behind the rocks, which will cause subsidence.

The final construction should have lots of tiers and planting pockets, and any flat tops on stones can provide easy walking access for weeding.

CHOOSING AND BUYING ROCK GARDEN PLANTS

Once you have a space to grow your plants, be it pots, a rock garden or raised bed, it is time to choose some plants. Most alpines for sale in garden centres, nurseries and online are tried-and-tested favourites for a wide variety of growing conditions, including being planted directly in the ground.

The main factors to consider when selecting alpines are their needs for light and water. Most will not tolerate being overshadowed or being waterlogged for long periods of time. Helpfully, growing them in a raised position (in pots, raised beds or a rock garden) allows for improved drainage

1. Build the foundation using hardcore/coarse grit at least 20cm/8cm thick.
2. Add a layer of hessian, to prevent soil from filling up the air gaps in the drainage layer.
3. Cover with a 30cm/12in layer of compost mix.
4. Decide where you want the rock layers to go (here white labels are used as markers). Using a trowel, dig spaces for the rocks one at a time, starting with the largest rocks at the bottom. Position them leaning back into the mound for stability.

5. Backfill with compost as you go and firm in well. Use the trowel to ensure there are no large air gaps behind the rocks that will cause subsidence.
6. The final construction should have lots of tiers and planting pockets.

There are not many specialist alpine cultivation tools, so you sometimes have to get creative. Repurposing a plastic bottle will make a great small scoop, as will using old cutlery for planting small alpines.

and better access to light. Generally, most rock-garden plants need at least a few hours of direct sunlight a day, preferably of morning sun.

Spring and early summer are the seasons when garden centres are overflowing with alpines in full bloom. Another good source is specialist alpine nurseries that grow plants with care — many also offer mail-order services. Often, small plants prove good value for money as they are cheaper than more established, pot-grown plants and may be more likely to survive long-term if treated correctly.

ROCK GARDENING EQUIPMENT

Many general garden tools, such as a spade, hand fork and watering can, are equally useful in an alpine context; however, there are some specialist tools for small plants. For example, you may prefer to use scissors or snips, as secateurs might be too large. Tweezers can be helpful for deadheading extremely small plants. A scoop for grit is a must-have, whether it is a purpose-bought one or a repurposed plastic water bottle cut in half. Sharp and narrow trowels are invaluable for reaching into tight gaps between rocks.

POTTING COMPOSTS, SOILS AND TOPDRESSINGS

When growing in a raised bed or rock garden
In such places, a soil-based compost such as John Innes No. 2 or No. 3 is ideal when mixed in equal parts with grit, to create a loose, free-draining mix that can retain moisture without being too wet. Soil-based compost also stores moisture and nutrients during summer. Avoid multipurpose potting compost as it can retain too much moisture and can promote too much lush, fast growth, creating plants that may be out of character.

An alternative growing media for alpines is pure horticultural sand; this can work well for a surprisingly wide variety of plants. Why not experiment to see what is best in your climate and region? Sand has the advantage of being cheaper than potting compost, and it always retains its structure, allowing water to pass through freely. It has the disadvantage of perhaps requiring more watering and feeding in summer, so may be more suitable for succulent or drought-tolerant plants.

Drainage is your friend when it comes to many alpine plants. There is a wide choice of potting media suitable for growing these plants: for example (here, starting top right and moving clockwise), LECA, fine pumice, large pumice, John Innes No. 3, perlite, small grit, vermiculite.

When growing in pots and troughs

In pots and troughs, alternatives to grit can include pumice, perlite, vermiculite, crushed brick, lightweight expanded clay aggregate (LECA) or clay granules (e.g. Seramis™). These all have the advantage of being lighter in weight and retaining more air in the growing media than grit or soil-based John Innes No. 3. They may be more useful for harder-to-grow plants, too, but they are more expensive and less commonly available. Therefore, grit may be the easiest potting media drainage option for many people. The airier you make a compost, the more you have to water the pot, or else think about utilizing a sand plunge (see Using a sand plunge for alpine cultivation, page 128).

Topdressings

Whether a plant is growing in the ground or in a pot, topdressing is an essential part of alpine cultivation. A layer of grit over the surface not only looks attractive, but it also helps to suppress weeds and retain moisture while keeping the neck of the plant above the soil layer below.

Choices of grit are entirely up to personal taste; however, it generally looks more cohesive to match the grit to the same stone used in your rock garden. Coarse grit (diameter 2cm/¾in) is great for keeping the neck of the plants really dry in an outdoor setting, as water passes through very quickly, whereas fine grit (0.5cm/¼in) is more ornamental in pots and will retain more moisture at the surface.

PLANTING

For most alpines, this is an activity best carried out in spring (see Planting a rock garden, page 92). Planting in pots and troughs is a similar process.

Topdress a pot or other container, here with some fine granite grit, ensuring it is under the foliage, here of houseleek (*Sempervivum*).

Always ensure the root ball of the plant is lightly teased apart before planting, and that each plant is positioned at the same depth or slightly higher than it was in its pot. Water in after planting. A notable exception are cushion-forming plants, in particular *Dionysia*, which do not enjoy too much root xdisturbance when planting.

Some succulents, such as agave, liveforever (*Dudleya*) and cacti, prefer to be left dry for a few days after planting, so wounds can callus to discourage rotting; this is specified in their individual plant profiles (see pages 42–133).

When watering alpines (here *Globularia* spp.) it is best to do so from the edge, to avoid wetting the centre of the plant, which encourages rot.

WATERING

Although alpines have specific adaptations to minimize water loss, these adaptations may be less effective at warmer, lower-altitude environments, when the plants are not growing up a mountain. Species with thickened taproots or tubers, lots of hairs or very dense growth may be prone to rot when overwatered, especially when watered from overhead. This can be avoided to some extent by watering at the base of the plant or the soil next to the plant, so the foliage remains dry.

FEEDING

Many alpine plants are specifically adapted to survive low-nutrient environments. This, however, does not mean they do not benefit from fertilizer.

For pot-cultivated plants, feeding regularly throughout the growing season with a low-nitrogen fertilizer, such as a liquid tomato one, will encourage flowering without promoting excessive leafy growth. Outdoor-grown plants in raised beds or rock gardens will benefit from the application of a balanced fertilizer such as a granular one, or blood, fish and bone meal in spring.

REPOTTING

The majority of alpine species have extensive root systems in comparison to their top growth. Although tolerant of being left undisturbed for a number of years, repotting encourages new growth and flowering, by supplying new nutrients. It also allows you to inspect the root health for pests and disease such as root aphid and vine weevil larvae (See Pests and diseases pages 134–7).

Like planting (see page 30), repotting is best done in spring, ideally just before flowering when plants are actively growing.

PRUNING

Simply removing excessive dead leaves, diseased stems, clearing up leaf litter and deadheading should be all you need to do to keep plants tidy and healthy. Because of the slow-growing nature and generally small stature of many alpines, they do not require frequent pruning unless they start encroaching on neighbouring plants. This is most common with mat-forming plants, which can be controlled by snipping stems just above a node (joint) around the edges, back to the position you desire. Pruning is best done when plants are actively growing and after flowering so as not to remove this year's blooms.

Repot a slow-growing alpine species (here globe daisy/*Globularia*) into a pot one size larger than its current one, to help avoid rotting.

Spent flowers are regularly removed in the Davies Alpine House at Kew. Plants such as cyclamen require frequent deadheading when grown under glass or in moist climates, to reduce the amount of mould formation on the dead stems.

This is not usually a problem when alpines are cultivated outdoors, where there is frequent air circulation.

HOLIDAY CARE

When leaving your alpines for a week or two to go on holiday, there are a few simple things you can do to ensure they remain healthy until your return.

Generally during winter, plants can be left alone unwatered for many months at a time without risk of drying out.

In summer, plants in the ground will be fine for quite a few weeks except during very hot and dry weather. It might therefore be worth asking a neighbour for help to water your outdoor plants every couple of weeks if you are going to be away for a month or longer.

Plants in containers can be more at risk of drying out if left longer than two weeks in warm weather. Make sure you water all containers thoroughly before you leave, and move pots to more shaded positions. You can also employ the use of shade netting or horticultural fleece, to cast shade over pots you cannot move; this can be purchased from your local garden centre or online. You can also use timer-drip irrigation systems especially made for pots that can be set to water your containers on a weekly or two-weekly basis.

Bulbs in pots require watering only during their growing season, which is usually between autumn and early summer, depending on the species. They can be left dry in a sheltered position in a cool shed or under a greenhouse bench for their dormancy period.

PROPAGATION

Propagating your own plants allows you to become much more closely acquainted with them and understand how they grow. It is also a great way to fill up empty gaps in the garden and share plants with friends. Always propagate from healthy new growth that is free of diseases such as viruses and of pests such as aphids and mealy bugs (see Pest and diseases, pages 134–7).

There are many ways to propagate alpine plants, but they are broadly in two main categories: vegetative propagation to produce clones (using cuttings, offsets/runners, layering and division; see pages 34, 39, 39 and 40, respectively); and seed propagation, which produces unique, non-clonal offspring (see page 40). If you have a specific cultivar such as a variegated form of a plant, you must use vegetative propagation, to ensure the cultivar features are retained.

Many true alpine species are very easy to increase by vegetative propagation because they are adapted to unstable changing habitats, where it is beneficial to be able to re-root downhill if a part of a plant is broken off by a landslide or severe wind.

For most alpines, when propagating vegetatively, a free-draining substrate is required for good rooting. This can comprise a variety of components, depending on the species and time of year. Some examples are pure horticultural sand, fine pumice, perlite or a mix of equal parts John Innes No. 1 and any of the aforementioned media. The main idea is that humidity is retained, but water freely passes through the substrate.

Having inserted each piece of propagated material into the substrate, leave them somewhere cool such as a shaded cold frame, to minimize water loss. Ensure they are kept evenly moist, but not overwatered: for example, in early spring, you may need to water only every week or two.

Cuttings

These are traditionally categorized into softwood (summer), semi-ripe (autumn) and hardwood (winter) cuttings, depending on the time of year in which they are ready to be taken and on the plant life form (see Alpine life forms explained, page 15). The rules are slightly more flexible with alpine species as they have different degrees of woodiness. There are a few special alpine variations on these traditional three categories, which are discussed in the following pages.

Although cuttings can be taken successfully at any time in the year, it is best to do this when the weather is not too hot (that is, in spring) or when the plant is actively growing (in autumn). This ensures that water loss is minimized from the cuttings. The exception is hardwood cuttings, which are taken in winter.

Always collect cutting material early in the morning or on a damp/overcast day; this means each cutting will be at its most hydrated. Using a clean knife or snips, remove non-flowering stems each with at least 3–4 nodes (points where the leaves or stems emerge from the main stem). Discard the leaves/stems from the bottom of each cutting, to expose at least two nodes; these will be buried in the cuttings mix. Placing cuttings around the edge of the pot allows you to separate them more easily once they root.

Vegetative propagation categories

Softwood cuttings

Phlox (*Phlox*)

Semi-ripe cuttings

Rock rose (*Helianthemum*)

Hardwood cuttings

Willow (*Salix*)

Basal cuttings/divisions

Auricula (*Primula*)

Rosette cuttings

Saxifrage (*Saxifraga*)

Succulent stem cuttings

Liveforever (*Dudleya*)

Offsets/Runners

Houseleek (*Sempervivum*)

Layering

Willow (*Salix*)

Division

Thyme (*Thymus*)

Simple polycarbonate cold frames are the perfect place to store your cuttings as they root. The protected space maintains constant humidity and regulates temperature.

Softwood cuttings are used for soft-growth propagation of non-woody material in summer using the current season's growth. This will work for genera such as phlox, pinks (*Dianthus*) and thrift (*Armeria*). Cut new-growth stem tips, 3–5cm/1¼–2in long, then carefully remove the lower leaves and insert at least two nodes of each cutting into the cuttings mix (see Propagation, page 34) in a pot.

Stem cuttings from most cushion-forming plants are considered 'softwood'. See also Propagating a cushion-forming plant on page 66.

Semi-ripe cuttings are suitable for species that have slightly woody stems, such as mountain avens (*Dryas octopetala*) or rock rose (*Helianthemum nummularium*). They are best taken in late summer, when the wood is hardening off. Snip off healthy stem tips, 5–10cm/2–4in long (depending on size of the species), with a harder base and softer tip. Cut off the lower leaves growing from the woody stem. Insert each cutting into the cuttings mix (see Propagation, page 34) in a pot, and place in a cool shaded cold frame. A heated mat or heated propagator, to provide warmth from below, can help to speed up the rooting process.

Hardwood cuttings are taken in winter and used if you have a deciduous shrub such as net-leaved willow (*Salix reticulata*) or mountain cherry (*Prunus prostrata*).

Collect some straight woody stems, 10cm/4in long, cut below a node, and then insert each cutting into a pot filled with cuttings mix (see Propagation, page 34). Keep the cuttings cool but frost-free over winter, ensuring they are moist but not sitting wet. You should notice roots forming by spring, usually before the buds burst on the stems.

Basal cuttings/divisions are very easy to propagate and can be from non-woody or slightly woody material, depending on the species. Essentially, they are miniature divisions from the base of a plant. This works for herbaceous perennial plants as well as for plants that have tightly clustered stems at the base. Look for a stem about 5cm/2in long that appears to be producing its own roots. Using a sharp knife or small trowel, sever this at the base and pot it up separately in cuttings mix (see Propagation, page 34). Keep it cool and shaded for a few weeks, to establish. This technique is good for auricula (*Primula auricula*), page 103.

Rosette cuttings are miniature softwood or semi-ripe cuttings. Many cushion-forming plants and mat-forming ones grow in rosettes. Taking cuttings of these can be tricky as the distance between leaves and stems is very short. The technique for selecting and removing rosettes is

Cuttings can be potted separately once rooted. This example shows some well-rooted semi-ripe cuttings from lavender-leaved sage (*Salvia officinalis* subsp. *lavandulifolia*), a small subshrub suitable for a rock garden.

described in Propagating a cushion-forming plant (page 66). Some rosette-forming species to propagate in this way include saxifrages (*Saxifraga*).

Some species will even form roots on almost all of the bases of their rosettes as they grow along the ground. This is true of houseleeks (*Sempervivum*); they can be more like a division than a true cutting.

Succulent stem cuttings are suitable for the cacti species mentioned in this book as well as for succulents such as liveforever (*Dudleya*) and *Lewisia*. This is the only stem-cutting technique that varies considerably from those described

above. Succulent species with thickened stems such as these are prone to rotting if inserted straight into wet potting soils after being severed from the parent plant. In summer, therefore, when the plant is actively growing, cut a stem, 5cm/2in long, from the parent plant with a very sharp knife. Leave this somewhere shaded and dry for a week or so. It requires this period for the wound to dry (callus). Insert the cutting into some moistened, well-drained potting media, such as 80 per cent perlite and 20 per cent John Innes No. 1. Place the pot in a part-shaded area and keep the potting media slightly moist but allow it to dry slightly between waterings.

Strawberry begonia is easy to propagate, as it produces abundant runners (aka stolons) in the growing season.

Houseleeks are one of the easiest alpine plants to propagate by division of offsets, and they can be kept going for generations of gardeners.

Offsets/runners

A few alpine species reproduce vegetatively in the wild using a variety of techniques that we can adopt to create more plants. Plants such as houseleeks (*Sempervivum*) regularly produce offsets as part of their life cycle after flowering in summer. These will produce their own roots quite quickly and can be teased away from the parent plant and replanted. Plants such as strawberry begonia (*Saxifraga stolonifera*) produce runners (stolons) in their growing season. You can leave them attached to the parent plant and they will root of their own accord into the surrounding soil given some time, or you can place a pot with some compost underneath the runner rosette and it will root in the pot ready to be severed after a few weeks.

Layering

Layering is another technique that mimics natural processes – many species layer themselves as debris builds up over stems. It is suitable for woody species such as willow (*Salix*) at any time of year but is best in spring, using ripe wood not fresh tip growth. Layering is also effective for plants such as garland flower (*Daphne cneorum*) that may be more difficult to propagate from cuttings.

Find a healthy stem at least 20cm/8in long and strip the leaves from the middle 10cm/4in. Bend this leafless section down

Growing alpines from seed is a rewarding way to produce many new plants. Seedlings – here, fringed pink (*Dianthus superbus*) – can be pricked out once at least two true leaves have formed above the seed leaves.

to the ground and weigh it in place with a pebble or secure with a hoop of wire. Bury this section under 5–10cm/2–4in of grit or soil, leaving the tip of the stem with leaves exposed. Wounding the bare stem by carefully scraping the surface of the bark to expose the green layer (cambium) in a 2.5cm/1in section of the downward-facing wood, and applying some rooting hormone, before burying it may speed up the rooting process. The layered area needs to be kept moist all year, to encourage rooting. It may take up to a year for the layer to root sufficiently for it to be severed from the main plant and potted up or replanted elsewhere.

Seed

Growing alpines from seed encourages healthy vigorous young plants by crossing the genetics of both the parents. It is not suitable for cultivars or variegated plants if you want to reliably retain their characters. Growing alpines from seed needs some patience; many species take at least two or three years to flower from seed-grown plants. See also Growing alpines from seed, page 60.

Division

This propagation technique is suitable for a wide range of evergreen and herbaceous perennial alpines that have multiple stems or buds attached to their own root systems. Some examples include trumpet gentian (*Gentiana acaulis*) and liverleaf (*Hepatica nobilis*). For a step-by-step guide on how to divide an herbaceous perennial alpine plant, see page 54.

Division is also an ideal method for propagating geophytes (for example, bulbs, tubers and corms). This is best done when they are dormant as they will be without roots: for example, plants such as tulips (*Tulipa*), fawn lilies (*Erythronium hendersonii*) or Bukhara iris (*Iris bucharica*) are best divided in late summer.

Once you have lifted a dormant geophyte, inspect it for pests (see Pests and diseases, page 134), then divide it and replant following the advice in Planting a patch of early squill bulbs (see page 118). You can leave a label or bamboo cane to remind you where the clump is.

Division of geophytes can also be done carefully in growth or 'in-the-green', especially if you are not certain where the bulbs/tubers/corms are when digging them up, so there is a risk of cutting through one by mistake. Snowdrops (*Galanthus*) are traditionally divided in-the-green and tolerate this well; ensure you replant to the same depth as previously, and water well after replanting.

Make sure you are careful with the delicate roots when dividing snowdrops (*Galanthus*) 'in-the-green'.

Plants

Hardy agave

Agave aka century plant, mountain agave

Once established, some species of these rosette-forming evergreen succulents from the Americas produce giant flower spikes, up to 7.5m/25ft high. Many species occur at very high altitudes in the mountains – not what you expect from an agave. Hardy agaves are great for adding spiky drama to a rock garden.

—

WHERE TO GROW

Grow in the sunniest, most well-drained spots in a rock garden or crevice garden (see Planting up an alpine crevice trough, page 48).

HOW TO GROW

Plant in mid-spring to give hardy agave time to establish well. Rosettes die after flowering, but offsets will often form and can be split off and replanted or potted up in spring (see page 39).

GROWING TIP

Although hardy agaves are relatively cold-hardy, down to −5°C/23°F if dry, they may require winter protection from wet (see Winter protection, alpine houses and sand plunges, page 22).

Agave parryi subsp. *neomexicana*

Family Asparagaceae	
Height & spread 50–100×50–100cm/ 20–39×20–39in	
Flowering time Late spring–summer	
Hardiness H3–H4	
Position Full sun	

TRIED AND TESTED
Some of the hardiest *Agave* species growing outdoors at the Royal Botanic Gardens, Kew, include *A. montana*, *A. americana*, *A. havardiana*, *A. filifera* and *A. striata*.

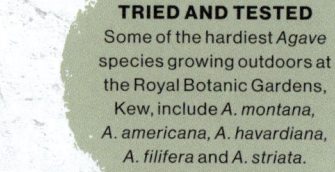

Agave filifera

Alpine lady's mantle

Alchemilla alpina

This no-fuss true alpine is the smaller cousin of the common lady's mantle (*A. mollis*). It is a mound-forming perennial with small, star-shaped leaves and clusters of tiny, yellow-green flowers, 15cm/6in high.

—

WHERE TO GROW

Alpine lady's mantle thrives in a moist open position in a rockery or large pot, but tolerates a wide variety of conditions.

HOW TO GROW

Plant in spring, adding organic material to the planting hole. Water in well and keep moist through the growing season. Alpine lady's mantle is very easy to propagate by lifting and dividing (see Dividing an herbaceous perennial alpine, page 54); do this once there is minimal risk of frost, in early or mid-spring.

GROWING TIP

This ground-cover plant may be good to pair with winter-flowering bulbs such as *Crocus* or squill (*Scilla*; see page 117) so you can enjoy two seasons of interest.

Family Rosaceae

Height & spread
15–20×50cm/
6–8×20in

Flowering time
Summer

Hardiness H6

Position Full sun–part shade

THE ALCHEMIST'S FRIEND
Raindrops stay on the leaves of lady's mantle due to the dense hairs. In folklore, it was believed that these leaves could be used by an alchemist for potions – hence the genus name *Alchemilla*.

Aloe

Aloe

For a vibrant look to your rockery and troughs introduce these evergreen perennials with their spiky leaves and red-hot-poker- (*Kniphofia-*) like flowers, 50cm/20in high. They belong to a group of succulents of mainly African origin – the species listed below being from the mountains of South Africa and Lesotho.

—

WHERE TO GROW

Aloes require bright light and well-drained conditions to thrive; they will not tolerate excessive moisture or hard frost in winter.

HOW TO GROW

When planted in a rockery, aloes may need protecting through winter (see Winter protection, alpine houses and sand plunges, page 22). If growing in a pot, bring them under cover for winter. Most species can be propagated by division in summer (see Dividing an herbaceous perennial alpine, page 54) or by seed (see Growing alpines from seed, page 60).

GROWING TIP

Aloes enjoy being planted on a vertical slope in a rock garden, crevice garden (see page 18) or a stone wall, enabling water to drain from their rosettes.

Family Asphodelaceae
Height & spread 15–75×10–50cm/ 6–30×4–20in
Flowering time Summer
Hardiness H3
Position Full sun

Aloe perfoliata

MATHEMATICAL PERFECTION

The foliage of the spiral aloe (*A. polyphylla*) grows at angles that follow a perfect mathematical Fibonacci sequence. This pattern can also be observed in galaxies, shells and other plants such as houseleek (*Sempervivum*).

Aloe polyphylla

NOTABLE SPECIES AND OTHER GENERA

- *A. perfoliata* (mitre aloe) has a creeping habit and red flowers; it can survive freezing temperatures if kept dry.
- *A. polyphylla* (spiral aloe), from the mountains of Lesotho, can tolerate some wet and snow.
- *Aloiampelos striatula* (striped-stemmed aloe) forms a multi-stemmed shrub with yellow flowers; nicknamed the 'hardy aloe'.
- *Aristaloe aristata* (lace aloe) bears orange flowers and spreads to form clumps of rosettes.

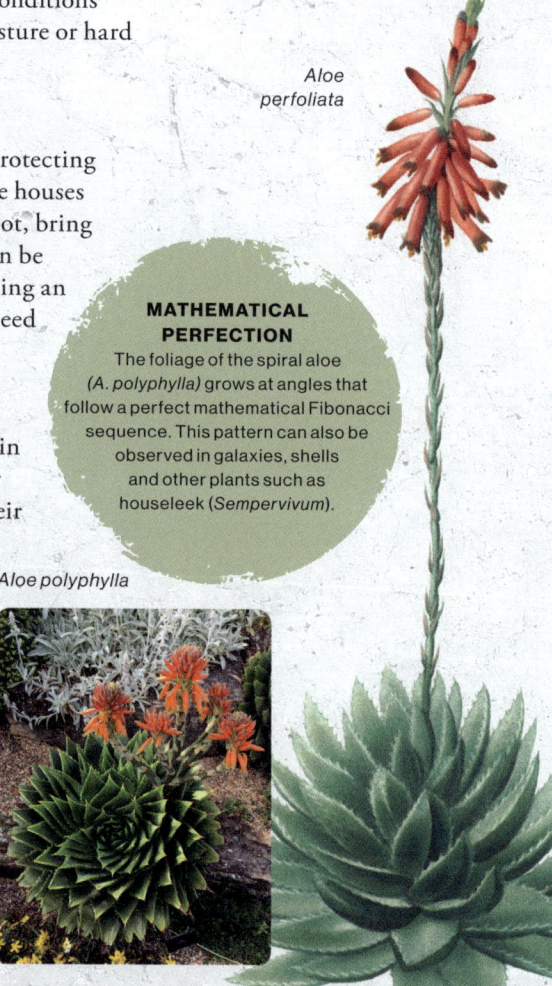

Shaggy rock jasmine

Androsace villosa aka fairy candelabra

This delicate Eurasian alpine plant forms a perfect cushion of dense rosettes, with its tiny leaves covered in fine hairs. The starry flowers grow in spring, just above the foliage, 2.5cm/1in high.

—

WHERE TO GROW

Grow outdoors where shaggy rock jasmine can get lots of light and air, in well-drained soil. It may need protection from winter wet (see Winter protection, alpine houses and sand plunges, page 22).

HOW TO GROW

Plant in spring in a crevice garden (see Planting up an alpine crevice trough, page 48) or a gritty rock garden or raised bed. Keep watered during the first season as well as through subsequent dry periods in summer. May need shading from hot sun in summer. Propagate by cuttings in spring (see Propagating a cushion-forming plant, page 66).

GROWING TIP

Being native to limestone cliffs and scree, shaggy rock jasmine prefers alkaline soil, but this is not essential, provided that you have added lots of grit to the planting area.

Family Primulaceae

Height & spread
2.5–5×5–20cm/
1–2x2–8in

Flowering time
Spring

Hardiness H5

Position Full sun

CHAMELEON FLOWERS
The flowers each start with a yellow eye in the centre but, as they are pollinated and begin to fade, the centre turns cerise, then gradually the whole flower goes a soft pink. This means there is a variety of flower colours on the same plant.

Planting up an alpine crevice trough

A trough or any container of your choice is the perfect way to experiment with the crevice garden concept (see page 18) on a small scale. You can also build your own trough (see Casting a hypertufa trough, page 100).

Apart from being practical, alpine crevice troughs are also quite attractive and mimic natural rock formations and the way that plants grow in their native habitat. They are also a good way to use up some crocks (broken pots), roofing tiles or slates. The flat surfaces of the erect pieces positioned in the trough provide a cool root run for the plants, encouraging deep rooting, while the small spaces between the crevices are ideal for growing very young plants such as cuttings and seedlings. The important thing to remember is that the upright pieces must be well-buried into the potting media so that the flat surfaces can function correctly.

When selecting plants to grow in your alpine crevice trough, bear in mind that plants in troughs can get tired after a few years, so why not seize the opportunity now to experiment and see what plants and potting media work best in your climate. You can then adapt the planting as appropriate later. (See also Potting composts, soils and topdressings, page 29.)

Before you start to add the potting media, the erect crocks, tiles or slates and the plants, position your trough where you want it to remain and raise it up on bricks. It is harder to move a trough once it is fully planted.

With regards to an appropriate potting media, pure horticultural sand is often used for crevice gardens on a large scale. However, in a small trough, the sand might dry out quite quickly. A 50:50 mix of grit and John Innes No. 3 should be a bit more moisture-retentive – and more convenient if you prefer not to water every few days in summer.

1. Cover the drainage holes with some crocks, fill the bottom 3cm/1¼in of the trough with coarse grit and then cover with hessian.
2. Put a thin layer of potting mix over the hessian. Starting at one end, add tiles vertically almost all the way to the bottom. Leave some planting gaps no larger than 2–5cm/¾–2in.
3. Fill in between the gaps with potting mix as you go, using a stick to work the mix well down into the trough.
4. Use a fork or spoon to dig out the planting pockets for your small plants (here, *Dudleya lanceolata*, *Sempervivum* 'Sirius', *Petrosedum sediforme* 'Gold' (aka *Sedum sediforme* 'Gold'), *Aristaloe aristata* and *Lewisia cotyledon*). Don't plant too densely.
5. Add a topdressing of grit, then water the plants well.

Alpine columbine

Aquilegia alpina aka alpine granny's bonnet, breath of God

This small herbaceous perennial is native to the European Alps, where it grows in limestone soils. It has small, blue-green, dissected foliage and downward-facing, purple flowers, 15cm/6in high.

—

WHERE TO GROW
Grow in a rockery or raised bed with appropriate drainage. Being native to alpine meadows, it enjoys bright conditions and even moisture levels; it doesn't like to dry out or sit in wet ground.

HOW TO GROW
When planting, add garden compost to keep the soil evenly moist. Being a short-lived plant, alpine columbine may die after a couple of years, but it will regrow from seeds sown from autumn to spring (see Growing alpines from seed, page 60).

GROWING TIP
To encourage new blooms, remove spent flowering stems. If you wish to promote self-seeding, instead retain the dead flowers while they form seeds.

Family Ranunculaceae	
Height & spread 20–30×10–15cm/ 8–12×4–6in	
Flowering time Early summer	
Hardiness H6	
Position Full sun–part shade	

TALONS OF AN EAGLE
The tubular hooked spurs on the backs of the flowers have evolved for long-tongued insects to source nectar. The talon-like shape of these spurs gives rise to the genus name *Aquilegia* – *aquila* meaning 'eagle' in Latin.

Juniper-leaved thrift

Armeria caespitosa aka *A. juniperifolia*

A smaller cousin of the common sea thrift
(*A. maritima*) is this little evergreen plant, which
forms a compact cushion of needle-like foliage. It
is covered in bright pink, pom-pom flowers on stems
10cm/4in tall, in spring.

—

WHERE TO GROW
This tough plant requires good gritty drainage and
full sun and is perfect for a rock garden or trough.

HOW TO GROW
Juniper-leaved thrift is very tolerant of drought
but will require watering to establish. It needs
minimal maintenance other than deadheading after
flowering. Propagate by seed (see Growing alpines
from seed, page 60) or by softwood cuttings in
summer (see Softwood cuttings, page 36).

GROWING TIP
Although it is tough, juniper-leaved thrift doesn't
tolerate competition well. Make sure it doesn't get
covered by leaves of other plants.

Family Plumbaginaceae

Height & spread
5x10–20cm/
2x4–8in

Flowering time
Spring

Hardiness H5

Position Full sun

ADAPTATION IS KEY
Thrifts (*Armeria*) are perfectly
adapted to the mountainous
and coastal areas they occupy
in their native habitats, where
they form cushions with thin
leaves and strong flower stems
to make them resistant to gale-
force winds.

OTHER NOTABLE SPECIES AND CULTIVARS
- *A. caespitosa* 'Bevan's Variety' bears particularly bright
  pink flowers.
- *A. maritima* (sea thrift) is widely spread across the
  world and is at least double in width to *A. caespitosa*;
  it's a tough hardy plant with pink flowers and is great
  for rock gardens.
- *A. pungens* (spiny thrift) is a Mediterranean thrift
  forming a small evergreen shrub, to 60cm/24in tall; it's
  good for rock gardens, but doesn't tolerate sustained
  wet and temperatures below −5°C/23°F.

Purple rock cress

Aubrieta deltoidea aka lilacbush

This tough ground cover hails from the mountains of the Mediterranean. It forms a large dense mat of slightly hairy, tiny rosettes smothered in four-petalled, pink flowers, 10cm/4in high.

—

WHERE TO GROW
Plant in a rockery, raised bed, wall or pot in full sun and with good drainage.

HOW TO GROW
Although it tolerates dry conditions when established, purple rock cress appreciates watering for the first few months. It can be cut back in the growing season if it is becoming too large. Propagate by cuttings in spring (see Propagating a cushion-forming plant, page 66). This evergreen perennial will also likely self-sow nearby.

GROWING TIP
Purple rock cress looks stunning when underplanted with grape hyacinths (*Muscari*; see page 96), which will grow through it and flower at the same time.

Family Brassicaceae

Height & spread
5–10×10–50cm/
2–4×4–20in

Flowering time
Late spring–summer

Hardiness H5

Position Full sun

NAMES CHANGE
The old name for the Brassicaceae family was Cruciferae, deriving from the Latin for 'cross-bearing', because most members of this family have four petals in a cross-formation.

Three-forked emerald carpet

Azorella trifurcata

This Patagonian carrot relative is a wonderful botanical curiosity: it creeps slowly along the surface of the ground to form an evergreen carpet. Tiny, yellow-green, umbel flowers form in summer.

—

WHERE TO GROW

Grow in a rock garden, raised bed or large trough, in full sun with plenty of added grit. Three-forked emerald carpet thrives with space to spread out slowly and is very sensitive to competition and overshadowing.

HOW TO GROW

Plant in spring, watering in well. It may need protection from excessive wet weather (see Winter protection, alpine houses and sand plunges, page 22). Propagate by softwood cuttings in summer (see page 36); keep young plants in pots until they are at least 5cm/2in wide, before planting in the open ground.

GROWING TIP

Three-forked emerald carpet looks great when planted between rocks, preferably in higher shelves of a rock garden or trough so its intricate beauty can be admired.

Family Apiaceae	
Height & spread 2.5–7x10–50cm/ 1–3x4–20in	
Flowering time Summer	
Hardiness H6	
Position Full sun	

A CUSHION CARROT?
Yes, that's right, this plant is in the same plant family (Apiaceae) as the humble carrot. Just like a carrot, it forms a thick taproot in search of moisture. Its rosettes of tiny, three-pronged leaves are much-reduced versions of a carrot's fluffy top.

Dividing an herbaceous perennial alpine

Many alpines and rock garden plants are small herbaceous perennials. You can tell if a plant fits this category by examining it closely to ensure that it has multiple soft growth points, each with its own roots. Such plants can be propagated by division, but this method is not suitable for plants that have only a single stem (for example, many cushion-forming plants and small shrubs).

Dividing a plant is a great way to get new free ones, and then to spread them out in the original planting spot or to move them to a new area. Smaller divisions can be potted up and given to a friend. Lifting the plant prior to dividing it helps you to check its root health and to examine it for any ground pests such as vine weevil larvae, root aphids and root mealybug (see Pest and diseases, pages 134–7). Division and replanting can also rejuvenate a tired old plant, kick-starting new growth and flowering.

The tools required are the same as if you were dividing a border herbaceous perennial – just smaller. Often a hand fork is sufficient to lift the plant, and hands are enough to split it. For larger plants, you can use two hand forks, back to back, to prise the clump apart.

Because you cause minor root damage when propagating a plant by division, this method is best undertaken when the weather is cool and humid – spring and autumn are ideal. It is essential to make sure that each plant is well watered before division, so it is fully hydrated.

1 Using a hand fork, lift the whole clump of the plant you want to divide (here, flower of Jove/*Lychnis flos-jovis*).
2 Find a natural gap in the middle of the plant and gently tease it apart.
3 Ensure that each piece that breaks away has new shoots and roots below; here, the original plant has been divided into three.
4 Replant the divided clumps at the same level as previously in the soil.
5 Refresh or apply topdressing, 3cm/1¼in deep. Then gently water in well to settle the soil underneath. Label each plant.

SOME PERENNIAL ALPINES THAT CAN BE DIVIDED

Alpine lady's mantle (*Alchemilla alpina*)
Creeping thyme (*Thymus praecox*)
Darwin's slipper (*Calceolaria uniflora*)
Italian bellflower (*Campanula fragilis*)
Lesser meadow rue (*Thalictrum minus*)
Orpheus flower (*Haberlea rhodopensis*)
Pygmy iris (*Iris pumila*)
Saxifrage (*Saxifraga*, some spp.)
Stonecrop (*Sedum*)
Trumpet gentian (*Gentiana acaulis*)
Viola (some spp.)
Zoys's bellflower (*Favratia zoysii*)

Darwin's slipper

Calceolaria uniflora

This evergreen perennial forms a mound of green foliage covered with pouch-like, dark-marked, yellow flowers, 10cm/4in high. The fleshy white lip on each pouch attracts birds as a food reward, thus pollinating it.

—

WHERE TO GROW

This plant is best in a cool, well-ventilated greenhouse or alpine house in free-draining, organic-rich, acid compost in a pot.

HOW TO GROW

Provide part shade, to stop leaf scorch, with a few hours of sun a day. Keep the compost barely moist in winter, but do not allow to dry out fully. Darwin's slipper is vulnerable to slug attack (see page 134). Propagate by division in spring (see Dividing an herbaceous perennial alpine, page 54).

GROWING TIP

This is a suitable candidate for cultivation in a sand plunge, which will keep the roots moist when required (see Using a sand plunge for alpine plant cultivation, page 128).

Family Calceolariaceae	
Height & spread 10×10–15cm/ 4×4–6in	
Flowering time Spring–summer	
Hardiness H4	
Position Part shade	

IN THE FOOTSTEPS OF DARWIN

These incredibly unusual flowers were collected by Charles Darwin in Patagonia while on the voyage of the *Beagle* in the 1830s. The herbarium at the Royal Botanic Gardens, Kew still holds Darwin's original dried pressed specimens.

Italian bellflower

Campanula fragilis

This stunning bellflower is native to the limestone mountains of central and southern Italy. Its tumbling bunches of pale purple flowers grow on stems as long as 30–40cm/12–16in from a woody rootstock.

WHERE TO GROW

Grow in a rock garden, crevice garden, wall or raised bed filled with sharp drainage material (see When growing in a raised bed or rock garden, page 29). Position small plants in crevices, to allow them space to tumble. The woody roots may rot in very cold, wet winters.

HOW TO GROW

Once established, Italian bellflower will be drought-tolerant. Cut back any dying flowering stems, to encourage more blooms. A big haircut is required in autumn, to keep it tidy. Propagate by seed in spring (see Growing alpines from seed, page 60) or by basal division from side shoots in spring (see Dividing an herbaceous perennial alpine, page 54).

GROWING TIP

Italian bellflower can be susceptible to slug damage in spring (see page 134), so check plants regularly and pick them off.

Family Campanulaceae	
Height & spread 15–30×30–50cm/ 6–12×12–20in	
Flowering time Summer	
Hardiness H4	
Position Full sun	

FRAGILE LITTLE BELLS

The genus name *Campanula* comes directly from Latin meaning 'little bell', while the species name *fragilis* means 'brittle', referring to the easily snapped stems.

OTHER NOTABLE SPECIES

- *C. persicifolia* (peach-leaved bellflower) is more upright and larger than *C. fragilis*; its flowering stalks reach to 1m/3ft tall and its leaves look like those on a peach (*Prunus persica*) tree.
- *C. portenschlagiana* (Dalmatian bellflower) is a very vigorous and hardy plant, spreading rapidly by seed and flowering profusely.
- *C. poscharskyana* (Serbian bellflower) is another very vigorous plant with a similar trailing habit to *C. portenschlagiana*.
- *C. rotundifolia* (common harebell) tolerates much colder, wetter weather than *C. fragilis*.
- *C. thyrsoides* (yellow bellflower) is a monocarpic biennial species with upright spikes of pale yellow flowers, to 1m/3ft tall.

Alpine clematis

Clematis alpina

The alpine clematis is native to the rocky slopes and mountain woodlands of Europe and Asia. It's a small- or medium-sized climber with dainty, pale purple, four-petalled flowers followed by fluffy seed heads.

—

WHERE TO GROW

This deciduous climber is suitable for only a large rock garden where it can sprawl out over the ground or climb up other shrubs. It needs moist soil and full sun or partial shade.

HOW TO GROW

Plant in spring after the last frost, adding in plenty of garden compost. Keep moist through dry spells. Propagate by semi-ripe cuttings in autumn (see page 36) or by seed sown fresh in autumn (see Growing alpines from seed, page 60).

GROWING TIP

Flowers are produced in spring on the previous season's stems. Prune these lightly after flowering, to keep alpine clematis in check.

Family Ranunculaceae

Height & spread
0.2–3x0.3-1.5m/
⅔–10x1–5ft

Flowering time
Spring

Hardiness H7

Position Full sun–part shade

CLIMBING UP THE MOUNTAIN

Alpine clematis is a true plant of the mountains, ever reaching for higher elevations. It uses its coiling, tendril-like leaf stems to clutch on to the branches of other shrubs in mountain woodlands at altitudes as high as 2,900m/9,500ft above sea level.

Sowbread

Cyclamen

This group of tuberous perennials hail mostly from the Mediterranean. They produce ornamental leaves from autumn to spring and flowers, 10cm/4in tall, with reflexed petals in shades of white and pink.

—

WHERE TO GROW

Many species tolerate deep shade but require good light to flower well. They are best in part shade in a rock garden or at the front of a bed.

HOW TO GROW

All species require moist but well-drained soil; this can be achieved with the addition of garden compost when planting. Propagate by seed collected in midsummer and sown fresh (see Growing alpines from seed, page 60).

GROWING TIP

Because the species flower at different times, you can have blooms from early autumn through to midsummer by planting a selection of sowbreads.

Family Primulaceae
Height & spread 5–15x15–30cm/ 2–6x6–12in
Flowering time Autumn–summer
Hardiness H3–H5
Position Part shade

INSECT CONNECTIONS

Cyclamen flowers are pollinated by bees, but require a buzz to release the pollen. The seeds are dispersed by ants, which are attracted to the sugar-rich coating called an elaiosome.

Cyclamen hederifolium

NOTABLE SPECIES

- *C. coum* (Eastern sowbread) is a Caucasian/Turkish species; it flowers in winter or early spring.
- *C. cyprium* (Cyprus cyclamen) is an autumn-flowering species from the mountains of Cyprus; it doesn't survive extended wet or temperatures below −5°C/23°F.
- *C. graecum* (Greek cyclamen) flowers in autumn and produces highly variable leaf patterning.
- *C. hederifolium* (ivy-leaved cyclamen) blooms in autumn before the beautifully patterned leaves appear.
- *C. purpurascens* (alpine or purple cyclamen) is an almost evergreen, summer-flowering species with sweetly scented blooms.

Growing alpines from seed

It is incredibly rewarding to grow alpines from seed. Many rock garden plants produce lots of seeds. Some of them will self-seed freely, while others may require a bit of extra help to germinate or establish before planting. Because the alpine season is quite early in the growing year, seeds can be ripe enough to collect as early as midsummer. Keep checking on your plants after flowering and watch for signs of ripening. The fruits often go dry and brown in colour when the seeds inside are ripe; and the seeds may turn from green to dark, or even shiny black.

If the plant from which you wish to extract seeds is in a pot, you can bring the whole plant to your potting area. If it is planted in the ground, cut off the whole flower stem and immediately put it in a paper bag so as not to lose any seeds. You then need to clean off all debris from the seeds. Sieves are useful for this – you can even use a colander for a low-tech solution.

A seed compost mix of two parts John Innes No. 1 and one part perlite makes a suitable light potting media that will prevent damping off and will subsequently aid the separation of seedling roots. Many species of alpines have very small seeds, so do not need to be sown deeply as this may prevent germination. Topdressing the seeds with a thin layer of grit while germination takes place helps to deter competitive weeds and retains moisture on the surface of the seed compost mix. Make sure you label your pot; also save spare seed in a labelled envelope to share or sow elsewhere.

Many hardy alpine species require natural stratification (a period of cold temperatures) to overcome dormancy and to encourage germination. Sowing in autumn and leaving the pots outside can encourage germination by spring. See also Seed, page 40.

1   Check the seeds (here, of *Barbarea rupicola*) are ripe by observing a change in colour of the fruits as well as the seeds themselves if the fruits have started to open.
2   Separate the seeds from the debris using a sieve.
3   Fill a pot with the seed compost mix up to about 2cm/¾in from the pot rim. Sow the seeds thinly on the seed mix.
4   Lightly cover the seeds with a little sieved seed compost mix and then grit. Label the pot.
5   Sit in shallow water to soak from below for a few hours. Then move the pot to a protected outdoor space, such as a cold frame, while germination takes place.

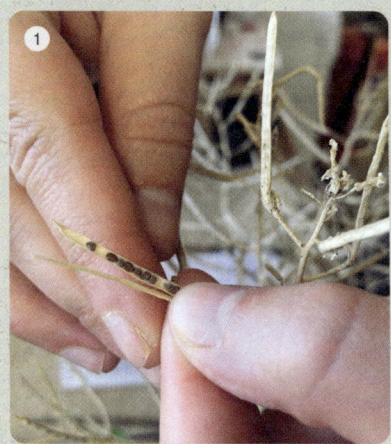

Yellow lady's slipper orchid

Cypripedium calceolus

The yellow lady's slipper orchid is a beautiful and rare European/Asian native terrestrial orchid. Its unusual flowers, each with burgundy petals and a yellow pouch, grow on stems about 50cm/20in tall.

WHERE TO GROW

Being a hardy, cool-growing species, yellow lady's slipper orchid requires a winter rest. It prefers slightly alkaline soil with plenty of organic material added when planting in spring.

HOW TO GROW

Grow in a part-shaded position at the front of a border/rock garden/raised bed. Water well after planting and during periods of drought. This slipper orchid can be susceptible to slug/snail damage (see Pests and diseases, page 134). Propagate by division of the roots in early spring (see Dividing an herbaceous perennial alpine, page 54).

GROWING TIP

Yellow lady's slipper orchid must be purchased from a reputable supplier to ensure that it has been propagated from cultivated stock.

Family Orchidaceae	
Height & spread 10–50x10–30cm/ 4–20x4–12in	
Flowering time Summer	
Hardiness H5	
Position Part shade	

A RARE SIGHT

Although more common in the European Alps, this species was down to just a single plant in the wild in the UK. Thanks to a technique known as micropropagation (with much work conducted by staff at the Royal Botanic Gardens, Kew), the numbers have been bolstered.

OTHER NOTABLE HARDY TERRESTRIAL ORCHID SPECIES

- *Bletilla striata* (Chinese ground orchid) is hardy (H4) and, in summer, has spikes, 50cm/20in tall, of bright pink blooms and long strappy leaves.
- *Calanthe striata,* a hardy (H3–H4) orchid from East Asia, produces spikes of fragrant flowers, 30cm/12in tall, followed by palm-like, pleated leaves.
- *Cypripedium formosanum* (Formosan lady's slipper orchid) is semi-hardy (H4) and bears white flowers each with a pink-mottled pouch.
- *Dactylorhiza fuchsii* (common spotted orchid) grows across Europe and Asia (H5–H6), preferring moist soil. It bears flower spikes, to 40cm/16in tall, in various shades of pink and white.
- *Pleione formosana* (Taiwan pleione) is hardy in mild climates (H3), and has storage organs called pseudobulbs. It needs very loose, organic-rich soils and part shade.

Garland flower

Daphne cneorum aka rose daphne

This delicate, southern European alpine beauty blooms in spring with highly scented, pink flowers. It is a low-growing, trailing subshrub, producing its flowers just (30cm/12in) above the evergreen foliage.

—

WHERE TO GROW
Although this species grows in mountainous regions, it is not reliably hardy in extended low temperatures (–5°C/23°F) in exposed conditions. Grow in full sun in moist but well-drained soil in a slightly sheltered position near a rock or wall.

HOW TO GROW
Add garden compost when planting, and water well. Once planted, garland flower is best not disturbed as it resents movement. Propagate by softwood cuttings in summer (see page 36) or by layering (see page 39).

GROWING TIP
Wash your hands after pruning or working with this plant – all parts are toxic.

Family Thymelaeaceae	
Height & spread 10–30x50–100cm/ 4–12x20–39in	
Flowering time Spring	
Hardiness H5	
Position Full sun	

A MYTHOLOGICAL NAME
In Greek mythology, Daphne was a water nymph who was transformed into a laurel tree to escape being ravaged by the god Apollo. The genus *Daphne* has leaves not too dissimilar to bay laurel (*Laurus nobilis*), hence its namesake.

Ice plant

Delosperma

This group of mainly South African succulents produces psychedelically bright, daisy-like flowers, 5cm/2in high, in all colours of the rainbow. The foliage is covered in bladder cells, giving a glistening icy appearance.

—

WHERE TO GROW

Plant in a rock garden, crevice garden, pot or trough, with 50 per cent grit. Ice plant requires the sunniest spot to thrive and flower well, and enjoys even moisture but not sitting in wet soil.

HOW TO GROW

Plant this evergreen succulent once there is minimal risk of frost, in spring, to maximize the summer growth period. Propagate by cuttings taken throughout the growing season (see Cuttings, page 34). Ice plant produces copious seed, and may self-seed.

GROWING TIP

Flowers open only in full sun, so don't be surprised to find a sudden lack of colour on an overcast day.

Delosperma dyeri Red Mountain

Family Aizoaceae

Height & spread
2–5×15–20cm/
1–2×6–8in

Flowering time
Summer

Hardiness H4

Position Full sun

WATER-ACTIVATED DISPERSAL
Once the seed capsules have formed and dried, they open only once wetted – a process known as hygrochasy.

Delosperma sutherlandii

Fringed pink

Dianthus superbus

Small mounds of grassy foliage give rise to 30cm/12in-tall flower spikes, crowned in summer with feathery-petalled, white or pink flowers with beautiful markings. This stunning perennial really packs a punch.

—

WHERE TO GROW

Although content in a wide variety of cultural conditions, fringed pink thrives best in a sunny, well-drained position in a rock garden or raised bed.

HOW TO GROW

Plant in spring and water well; it is drought-tolerant once established. Propagate from seeds sown fresh in autumn (see Growing alpines from seed, page 60) or by dividing plants (see Dividing an herbaceous alpine plant, page 54).

GROWING TIP

Deadhead whole flowering stems back to ground level in summer, to encourage more blooms.

Family Caryophyllaceae

Height & spread
10–30x20–30cm/
4–12x8–12in

Flowering time
Summer

Hardiness H4–H5

Position Full sun

A WIDESPREAD BEAUTY

Fringed pink has an incredibly large distribution in the wild, from western Spain all the way across Eurasia to Japan. There are six subspecies across its range, and these vary massively in colours, patterns and stature.

OTHER NOTABLE SPECIES

- *D. alpinus* (alpine pink) is a small, cushion-forming species with solitary large (4cm/1½in) flowers borne just above the surface of the cushion.
- *D. carthusianorum* (Carthusian pink) is a European herbaceous perennial with bright pink flowers in clusters on the ends of 60cm/24in-long stems in summer.
- *D. chinensis* (Chinese or rainbow pink) is a very-easy-to-cultivate herbaceous perennial, flowering on stems 30–50cm/12–20in long, in various shades of white and pink.
- *D. gratianopolitanus* (Cheddar pink) is native to the Cheddar Gorge in Somerset, UK, as well as in the rest of Europe. It forms very wide foliage mats, to 50cm/20in, and is smothered with pink scented flowers, 15cm/6in high, in early summer.
- *D. monspessulanus* (fringed pink) is longer-lived and forms wider cushions than *D. superbus*. The flowers are also slightly smaller and not as frilly.
- *D. webbianus* (aka *D. erinaceus;* hedgehog pink) forms very tight, dense, spiky cushions of evergreen foliage, to 50cm/20in wide; short-stemmed, pink flowers develop right on the cushion.

Propagating a cushion-forming plant

Many cushion-forming plants, such as *Dionysia* or saxifrage (*Saxifraga*), grow like miniature trees: they have a single trunk in the centre and branch outwards from there. Unlike trees, many species of cushion-forming plants retain all their dead foliage, accumulating this inside the cushion, where it forms a dense mass that acts as a thermal insulator. This can make taking cuttings tricky.

The best method of propagation is to lift carefully a lower edge of the plant in order to access single rosettes or small groups of rosettes. These can be cleaned up and inserted into a very free-draining media such as fine pumice, or else into a potting mix of at least 60 per cent perlite and the rest of John Innes No. 3 compost.

The key to successful propagation is to retain enough moisture to encourage rooting, but not keeping the cuttings mix so wet that the plant rots. It is much easier to achieve this balance in spring, when the sun is weak and the air is cool and moist. This way you will barely (if at all) have to water the cuttings until roots show through the holes in the bottom of the pots, after a few months. They are then ready to be potted up individually.

1 Lift the edges of the plant (here, *Dionysia aretioides*) and carefully snip out fresh healthy rosettes with stems at least 2cm/¾in long, using clean scissors.
2 Remove the lower dead leaves and cut the base of each stem neatly with a clean scalpel or scissors.
3 Use a stick or dibber to make evenly spaced holes for the cuttings around the edge of the pot filled with your cuttings mix. Insert the cuttings.
4 Label the cuttings and place them on wet sand somewhere cool, bright and humid but out of direct sunlight – a shaded cold frame or garden propagator is good.
5 Once rooted, pot plants up individually into a free-draining mix (see Potting composts, soils and topdressings, page 29). Then topdress and water.

Dionysia

Dionysia

This genus of magically miniature cushion plants is native to the mountains of central Asia. Flowers develop on the surface of their cushions, to 10cm/4in high, while leaves are only 5mm/¼in long.

WHERE TO GROW
Because dionysias grow naturally on exposed cool cliffs, in cultivation they prefer growing on tufa or in a part-shaded, alpine-house sand plunge in a pot (see Planting into tufa, page 112, and Using a sand plunge for alpine plant cultivation, page 128).

HOW TO GROW
Plant in growing media that provides lots of drainage: 80 per cent pumice, LECA, grit or perlite; and 20 per cent soil-based compost such as John Innes No. 2. Water the pots weekly only in spring/summer; keep the sand plunge just moist in autumn/winter. Propagate by cuttings in spring (see Propagating a cushion-forming plant, page 66).

GROWING TIP
To avoid wetting the foliage and causing the growth of fungus, water around the edges of a pot filled with dionysia.

Family Primulaceae

Height & spread
5–10x5–30cm/
2–4x2–12in

Flowering time
Early spring

Hardiness H3

Position Part shade

YOU'RE THE PIN TO MY THRUM
Like other members of the primula (Primulaceae) family, dionysia has two different flower types: the 'pin' form, where the stigma (female flower part) protrudes out of the flower; and the 'thrum' form, where the stamens (male flower parts) protrude out of the flower. Having a different flower type on separate plants means dionysia does not self-pollinate, thereby preventing inbreeding.

Dionysia curviflora

NOTABLE SPECIES
- *D. aretioides*, native to northern Iran, is one of the larger species in the genus; it bears yellow flowers and its leaves are covered in woolly hairs.
- *D. curviflora* is very compact, with single pink flowers; it comes from the cliffs of central Iran.
- *D. involucrata* is a larger-leaved species from central Asia (Tajikistan); it has branching inflorescences topped with pink flowers.
- *D. tapetodes* is compact, bears yellow flowers and has a wide distribution across the central Asian mountains.

Whitlow grass

Draba

These mat-/cushion-forming perennials and annuals hail from the temperate regions of the northern hemisphere as well as from South America. In spring, they form flower stems no more than 10cm/4in high, bearing four-petalled blooms, generally in white or yellow.

—

WHERE TO GROW
With 300 species to choose from, whitlow grass can vary in its requirements. However, all species need sun and good drainage (with at least 50 per cent potting grit added to the soil) and are good for a rock garden, raised bed, pot or trough.

HOW TO GROW
Some species may require winter protection from wet in an alpine house (see Winter protection, alpine houses and sand plunges, page 22). Propagate from seeds sown in spring (see Growing alpines from seed, page 60) or by rosette cuttings in summer (see page 37).

GROWING TIP
Many species of whitlow grass come from limestone-rich habitats in the wild so they may benefit from dolomitic lime added to their compost.

Family Brassicaceae	
Height & spread 2.5–10x5–20cm/ 1–4x2–8in	
Flowering time Spring	
Hardiness H5	
Position Full sun	

ALPINE CABBAGES
This genus of plants is in the same family as broccoli and cabbage – the Brassicaceae. There are many other alpine and rock-dwelling genera in this family including the rock cresses (*Aubrieta*) and wallflowers (*Erysimum*).

Draba aizoides

Draba aizoides

Mountain avens

Dryas octopetala aka eight-petal mountain avens, white-flowered dryad, white dryas

Iconic mountain avens is an evergreen dwarf shrub found in tundra and mountainous regions across the northern hemisphere. Its eight-petalled, white flowers form just above the foliage, followed by fluffy seed heads, to 10cm/4in tall.

WHERE TO GROW

Mountain avens is suitable for a rock garden or raised bed in well-drained soil. It does best in full sun in cool climates, and does not tolerate temperatures higher than 30°C/86°F for extended periods.

HOW TO GROW

Plant in spring, and water in well. Ensure mountain avens is kept from drying out in the growing season during warm spells. Propagate by semi-ripe cuttings in late summer or autumn (see page 36) or by seed sown fresh when ripe in early summer (see Growing alpines from seed, page 60).

GROWING TIP

Although this plant can be grown in a pot, it much prefers the cool deep root run that only the ground can provide.

Family	Rosaceae
Height & spread	0.05x2m/¹/₆x7ft
Flowering time	Spring
Hardiness	H7
Position	Full sun

A PLANT OF THE AGES
The time periods 'Younger Dryas', 'Older Dryas' and 'Oldest Dryas' (between 12,000 and 18,000 years ago) are named after *Dryas octopetala*. These were periods of global cooling, so this plant was abundant in the tundra landscape, where its pollen is found in vast quantities in the layers of soil cores.

Liveforever

Dudleya

This group of evergreen succulents comes from coastal cliffs and mountains in western North America. Liveforever produces stunning rosettes of silvery foliage, often bearing very brightly coloured flowers, on stems up to 15cm/6in.

—

WHERE TO GROW

Grow in a pot or trough, or in a rockery if your winters are not too wet. The key to successful cultivation is gritty drainage and good light.

HOW TO GROW

Plant at an angle to let water drain from the centre of the rosette. Some species have summer dormancy in the wild, so water very sparingly to minimize the risk of rotting. Winter protection will also help to keep liveforever dry (see Winter protection, alpine houses and sand plunges, page 22). Propagate by succulent stem cuttings in spring (see page 38); always allow the wound to dry for a few days before inserting the cutting.

GROWING TIP

Why not try growing liveforever in a pot with other succulents or cacti, for an interesting, low-maintenance display.

Family Crassulaceae	
Height & spread 10x15cm/4x6in	
Flowering time Summer	
Hardiness H3–H4	
Position Full sun	

NATURAL SUNSCREEN

Dudleya are protected from ultraviolet light damage by a waxy layer known as 'farina'. One species – *D. brittonii* – was found to reflect more ultraviolet than any other plant species, up to 83 per cent UV-B.

Dudleya cespitosa

Dudleya lanceolata

NOTABLE SPECIES

- *D. cespitosa* (coast dudleya) produces yellow flowers on stalks to 15cm/6in long.
- *D. cymosa* (canyon liveforever) is endemic to California and bears clusters of orange flowers.
- *D. farinosa* (powdery liveforever) has yellow flowers and blue-green leaves tipped with red when grown in full sun.
- *D. lanceolata* (lanceleaf liveforever) has thinner leaves with pointed tips.

Fairy foxglove

Erinus alpinus aka alpine balsam, starflower

A common sight in alpine gardens is this miniature perennial, which loves growing in rocky places. Fairy foxglove develops into a tight mound of hairy leaves, enhanced by bunches of pink flowers, 15cm/6in high, through the growing season. It is a good starter tufa plant (see Rocks, rocks, rocks, page 24), being drought-tolerant and alkali-loving.

—

WHERE TO GROW
Plant in any well-drained, sunny position in spring, and water in well. Minimal water is required when established.

HOW TO GROW
Propagate from seed (see Growing alpines from seed, page 60). Sow fresh seed on to pre-soaked tufa, then keep moist (see Planting into tufa, page 112).

GROWING TIP
Deadhead regularly with scissors, to encourage repeat-flowering and to discourage excessive self-seeding (fairy foxgloves will pop up everywhere).

Family Plantaginaceae

Height & spread
10–15×5–10cm/
4–6×2–4in

Flowering time
Spring–summer

Hardiness H6

Position Full sun

A TINY ALIEN
Although native to mainland Europe, the fairy foxglove can be commonly found in urban environments in the UK; it grows well between paving slabs and in lime mortar.

Henderson's fawn lily

Erythronium hendersonii

Nodding flowers, with pale pink, reflexed petals, are borne 30cm/12in high in early spring above mottled green foliage on this beautiful bulbous species. It occurs in the Siskiyou Mountains' woodlands of Oregon and California, USA.

—

WHERE TO GROW

Grow in moist, part-shaded, rich soil, in a raised bed or rock garden.

HOW TO GROW

In autumn, plant bulbs 5–10cm/2–4in below the soil surface, adding a handful of garden compost to the planting area (see Planting a patch of early squill bulbs, page 118). Propagate by lifting bulb clumps in summer, when dormant (see Division, page 40) and replanting elsewhere.

GROWING TIP

Henderson's fawn lily will diminish quickly if its foliage is covered by evergreen shrubs or leaves during the growing season.

Family Liliaceae

Height & spread 30x30cm/12x12in

Flowering time Early spring

Hardiness H5

Position Part shade

AN INTRODUCTION TO DISTRIBUTION
There is only one species of *Erythronium* native to Europe – dog's tooth violet (*E. dens-canis*); five species are from Asia while the other fifteen or so species occur in North America, including the Henderson's fawn lily.

Crimson bromeliad

Fascicularia bicolor

This hardy pineapple relative will bring a taste of the Patagonian mountains of Chile into your garden. Its striking blue flowers surrounded by crimson-red bracts are held within its long-barbed foliage.

—

WHERE TO GROW

Plant directly in the ground or in a large pot in well-drained soil and full sun.

HOW TO GROW

Crimson bromeliad is very easy to care for, with no pruning or maintenance required, although it may benefit from the addition of organic matter to the soil and extra watering in dry summer periods. Lift and divide in spring (see Dividing an herbaceous perennial alpine, page 54), taking care to avoid the sharp-edged foliage.

GROWING TIP

Covering with some horticultural fleece may help prevent ice damage when temperatures drop below –5°C/23°F.

Family	Bromeliaceae
Height & spread	50×50–100cm/ 20×20–39in
Flowering time	Summer
Hardiness	H4
Position	Full sun

BROMELIAD BONANZA
The bromeliad family (Bromeliaceae) contains many familiar tropical plants including air plants (*Tillandsia*) and pineapples (*Ananas comosus*).

Zoys's bellflower

Favratia zoysii aka *Campanula zoysii*

This unusual bellflower is endemic to the limestone mountains between Austria, Italy and Slovenia, where it is especially symbolic in the Julian Alps. Its small tufts of foliage are covered with narrow-ended, purple bells on erect stems, 10cm/4in high.

—

WHERE TO GROW
This slightly delicate, short-lived herbaceous perennial requires excellent drainage and is best grown in a pot or in an alpine house. It enjoys a bright position with no competition; in the wild, it grows on north/east-facing rock cracks.

HOW TO GROW
It requires cool conditions and doesn't tolerate winter wet. Although Zoys's bellflower can be slightly fussy in its growing conditions, it is easy enough to propagate by division in spring (see Dividing an herbaceous perennial alpine, page 54).

GROWING TIP
This plant is a great candidate to try on tufa, mimicking its natural growth habit on limestone rocks (see Planting into tufa, page 112).

Family Campanulaceae

Height & spread
5–10×10–15cm/
2–4×4–6in

Flowering time
Summer

Hardiness H4

Position Part shade

NECTAR THIEVES
Although pollinators can get into the end of each narrow bell bloom on Zoys's bellflower, short-tongued bees steal nectar by biting a window into the base of the flower.

Making an alpine cut-flower display

There is a long tradition of competitive miniature floral displays of cut alpine flowers at spring alpine shows. These often use spring-flowering bulbs such as snowdrops (*Galanthus*) and crocuses as well as herbaceous plants including hellebores (*Helleborus*) and shrubs like daphnes. If you are exhibiting at a show, it's a good way to show off the diversity of plants you've grown, but an alpine cut-flower display can also be invaluable to brighten your house in early spring. Another advantage is this doesn't have to be done just in spring, especially if your rockery contains a wide range of plants, as you will get interesting flowers and foliage year-round. Growing and cutting alpine flowers can be a good way to learn all the different ones in your garden and the times they bloom, as well as understanding what colours and textures work well together.

Mini 'posies', such as the one shown here, can be displayed in any small vase, cup or mug; you can even arrange one in something as small as a shot glass. Another option is to support the stems in a piece of florist foam, or in a ball of moss, placed in the bottom of a dish of water.

Think about plants that are strong and will help to hold the display, and those that are tall and more transparent to add airy height. Consider if you want the display to be symmetrical and viewed from multiple angles, or from just one side. A good way to build the floral display is to work from the edges up to the centre. The most important thing is to enjoy your plants and spend some time being creative.

1. Cut flowers with a long piece of stem (here, *Limonium delicatulum*) early in the morning to make sure they last well.
2. Immediately put the bases of the stems into some water.
3. Lay out your flowers into groups, to make it easier to decide how you will arrange them.
4. Build the shape of the display a little at a time, adding taller flowers into the centre of the arrangement if it is to be viewed from all sides.
5. Move your finished display to a position out of direct sun indoors, to keep it fresher for longer.

Elwes's snowdrop

Galanthus elwesii aka greater snowdrop

This common garden plant is native to the woodlands and mountainsides of the Caucasus region of Eurasia. Iconic, white, pendulous flowers, to 20cm/8in tall, form in winter.

—

WHERE TO GROW

Plant in soil with added garden compost to keep bulbs moist through summer. Find a spot in part shade at front of a rock garden or raised bed.

HOW TO GROW

In spring, purchase snowdrops 'in the green', growing in pots, for planting in naturalistic groups. Feed with a sprinkle of granular fertilizer once or twice in winter/ spring. After flowering, do not cut back the foliage until it's completely brown. Bulbs are easily separated from their clump when dormant in summer/autumn to spread around and can also be carefully lifted and divided 'in the green' in spring (see Division, page 40).

GROWING TIP

Watch out for viruses (see Pests and diseases, page 135) and narcissus bulb fly (see page 135), which commonly attack this plant's family.

Family Amaryllidaceae

Height & spread
20x30cm/8x12in

Flowering time
Winter

Hardiness H4

Position Part shade

A DRUG FOR ALZHEIMER'S
The bulbs of snowdrops and other plants in the daffodil family (Amaryllidaceae) contain an alkaloid chemical called galantamine. This is used in the treatment of Alzheimer's disease.

Checkerberry

Gaultheria procumbens aka American wintergreen, boxberry, partridge berry

Native to North America is this evergreen spreading shrub, which bears white bell flowers in summer. Its primary appeals are the subsequent bright red berries and its winter leaf colour.

—

WHERE TO GROW
Grow in part shade in a rock garden or raised bed in acidic soil with added garden compost.

HOW TO GROW
Keep moist through dry spells. If pruning is required, thin by removing stems down to the ground. Although checkerberry is a shrub, it suckers at the base freely, so propagate by division in early spring (see Dividing an herbaceous perennial alpine, page 54).

GROWING TIP
Checkerberry can spread quite vigorously when established, so don't plant it too near delicate or slow-growing plants.

Family Ericaceae

Height & spread
10–20x20–100cm/
4–8x8–39in

Flowering time
Summer

Hardiness H5

Position Part shade

NO BERRIES?
Gaultheria are dioecious, which means there are separate male and female plants. To guarantee a female plant, buy a plant with fruits forming on it, in summer.

Gaultheria cuneata

Trumpet gentian

Gentiana acaulis aka stemless gentian

Being one of the largest, single-flowered gentian species, trumpet gentian is a truly iconic alpine plant. Its small, shiny, dark green leaves form a mat, with electric, almost unnaturally blue, trumpet-shaped flowers, 10cm/4in high, at ground level.

—

WHERE TO GROW

This true alpine enjoys cool and slightly damp conditions, and is good for rockery cultivation in sunny, evenly moist, preferably alkaline, open positions. Make sure it isn't outcompeted by surrounding plants.

HOW TO GROW

Plant with added organic material in spring and keep well-watered throughout summer. Propagate by division in spring (see Dividing an herbaceous perennial alpine, page 54).

GROWING TIP

Trumpet gentian benefits by being shaded from intense hot sun in summer, so site it somewhere there is morning sun.

Family Gentianaceae	
Height & spread 5–10×10–30cm/ 2–4×4–12in	
Flowering time Late spring–summer	
Hardiness H5	
Position Full sun	

GETTING TO THE ROOT
Gentians host mycorrhizae (friendly fungi) in their roots. This is common in alpine plants and gives them an advantage when accessing nutrients in the poor alpine soils.

OTHER NOTABLE SPECIES

- *G. asclepiadea* (willow gentian) is a larger herbaceous plant than the trumpet gentian and is suitable for rich shaded soils.
- *G. clusii* (Clusius's gentian) looks very similar to the trumpet gentian and has similar cultivation needs.
- *G. lutea* (great yellow gentian) is another large, yellow-flowered species from moist European alpine meadows.
- *G. septemfida* (crested gentian) bears clusters of blue flowers and needs similar cultivation conditions to the trumpet gentian.
- *G. verna* (spring gentian) is a small-flowered, ephemeral species with similar cultivation needs to the trumpet gentian.

Globe daisy

Globularia vulgaris

This hardy, evergreen, mat-forming plant is native to
Europe. It has bright blue, tightly clustered flowers,
15cm/6in high, forming globes all over the mat.

—

WHERE TO GROW

As globe daisy thrives in bright, well-drained
conditions, it is an ideal candidate for the front edge
of a rockery.

HOW TO GROW

Plant in spring or autumn. Snip off the spent
flowers at the end of the season to keep the plant
tidy. Propagate by softwood cuttings in spring
(see Propagating a cushion-forming plant,
page 66), by division in spring (see Dividing an
herbaceous perennial alpine, page 54) or by seed
collected in late summer (see Growing alpines
from seed, page 60).

GROWING TIP

Deadheading through the flowering season
can encourage longer blooming.

<table>
<tr><td>Family Plantaginaceae</td></tr>
<tr><td>Height & spread
5–15×15–50cm/
2–6×6–20in</td></tr>
<tr><td>Flowering time
Summer</td></tr>
<tr><td>Hardiness H5</td></tr>
<tr><td>Position Full sun–part shade</td></tr>
</table>

'DAISY, DAISY...'
Despite its common name,
globe daisy is not a true
daisy. In fact, it's more
closely related to foxgloves
(*Digitalis*) and snapdragons
(*Antirrhinum*) in the
Plantaginaceae family.

OTHER NOTABLE SPECIES

- *G. cordifolia* (heart-leaved globe daisy) has a slower growth rate and more compact habit than *G. vulgaris*.
- *G. nudicaulis* (naked-stalked globe daisy) produces large shiny leaves and leafless flower stems.
- *G. repens* (creeping globe daisy) is one of the smallest of the genus, and grows on limestone rocks in the wild; its flowers are only 3cm/1¼in high.

Orpheus flower

Haberlea rhodopensis aka resurrection plant

This plant is found in the Rhodope Mountains of Bulgaria
and Greece, where it produces evergreen rosettes of dark
green, hairy leaves and showy, pale purple flowers, 10cm/4in
high, with orange-speckled throats.

—

WHERE TO GROW

Plant in a pot or trough, or in a rock garden. Although
drought-tolerant, Orpheus flower is best grown in a shady
moist spot out of hot afternoon sun.

HOW TO GROW

Plant in spring or autumn, adding plenty of organic material.
Water well and keep Orpheus flower topped up with water in
summer to ensure it looks its best; do not allow it to sit in
wet soil, however. Propagate by division of rosettes in spring
(see Dividing an herbaceous perennial alpine, page 54).

GROWING TIP

Try planting Orpheus flower in an angled crevice
of a rockery, to mimic growing on vertical rocks
in the wild.

Family Gesneriaceae

Height & spread
10–15×10–30cm/
4–6×4–12in

Flowering time
Spring

Hardiness H4

Position Part shade

DRIED TO A CRISP
The remarkable cellular
mechanism of desiccation
tolerance of this species has
been long studied. It has been
documented to resurrect after
three years of being totally dry –
hence one of its
common names.

Rock rose

Helianthemum nummularium aka flower of the sun

Rocky slopes and hillsides in Europe are the preferred habitat for this tough evergreen shrub. Its sunny yellow flowers continue through most of the growing season, forming just above the foliage.

—

WHERE TO GROW

Rock rose is a great choice for a dry sunny position in a raised bed, rock garden, pot or trough in well-drained soil.

HOW TO GROW

Although drought-tolerant when established, rock rose still needs regular watering when growing in a container. Deadhead after flowering, to encourage more blooms. Propagate by layering (see page 39), softwood cuttings in summer (see page 36) or semi-ripe cuttings in autumn (see page 36) or by seed sown fresh in late summer (see Growing alpines from seed, page 60).

GROWING TIP

This plant can get straggly with age, so thin out dead and old stems in summer, to make way for new growth from the base.

Family Cistaceae

Height & spread
10–15x30–50cm/
4–6x12–20in

Flowering time
Spring–summer

Hardiness H5

Position Full sun

FEELING YELLOW?
If yellow is not your favoured colour, there are a plenty of non-yellow-flowering cultivars of the common rock rose, including pink, red, orange and white ones.

Adding a small nature pond to a rockery

Alpine and mountain habitats always have water somewhere, whether that be a mountain stream, lake, pond or wet meadow. Such environments are often the most biologically rich part of the mountain. To emulate such habitats, why not add aquatic plants and water to your garden rockery, to make it a more nature-friendly, diverse and interesting space? After all, traditional rock gardens often have streams, ponds and boggy areas. For a small garden, you could purchase a preformed pond or large bucket from your local garden centre, or you could use a pond liner in a suitable hole in your rockery.

Plant choices are particularly important for a small pond. Aquatics grow very quickly, so fewer plants are better in this context – avoid the temptation to add too many plants. A well-behaved plant choice is a dwarf water lily such as pygmy water lily (*Nymphaea tetragona*), which will grow slowly at the surface of the water without casting shade on the plants around the edges. Choosing a site that gets at least a few hours of direct sun per day will encourage the water lily to thrive and flower well. It can be added to the pond in an aquatic basket, slightly raised from the bottom using bricks. Aquatic baskets contain the plants and allow you to easily remove them.

Another suitable plant could be hare's tail cotton grass (*Eriophorum vaginatum*). This plant is a common sight in marginal aquatic environments across the northern hemisphere including mountainous, arctic and boreal bogs and wetlands. It can be kept in an aquatic basket partially submerged into the pond and raised up on bricks.

1 Dig a hole that is slightly larger than your pond container (here, a large bucket) and line its bottom with 3cm/1¼in of builders' sharp sand to provide an even base.
2 Put the bucket in the hole and adjust the position with a spirit level until it is flat.
3 Backfill around the bucket edges with soil, tamping as you go.
4 Add a few bricks in the bottom of the bucket, to act as a podium for your aquatic plant. Fill the bucket with (preferably rain-) water.
5 Position the aquatic plant of your choice (here, *Nymphaea tetragona*) on the bricks, in its aquatic basket.
6 Position rocks and plants around the bucket edges to hide its rim.

SOME AQUATIC PLANTS FOR A SMALL ROCKERY POND

Blue flag (*Iris versicolor*)
Bogbean (*Menyanthes trifoliata*)
Common cotton grass (*Eriophorum angustifolium*)
Dwarf bullrush (*Typha minima*)
Frogbit (*Hydrocharis morsus-ranae*)
Hare's tail cotton grass (*Eriophorum vaginatum*)
Little floating heart (*Nymphoides cordata*)
Marsh marigold (*Caltha palustris*)
Pygmy water lily (*Nymphaea tetragona*)
Rigid hornwort (*Ceratophyllum demersum*)

Liverleaf

Hepatica nobilis aka common hepatica, kidney wort

Liverleaf is native to Europe and Siberia. The dainty, buttercup-shaped flowers, 10cm/4in tall, emerge in shades of pink and purple, before the variably marked, three-lobed leaves.

—

WHERE TO GROW
Grow in part shade in soil enriched with organic matter in the front of a rock garden, raised bed or container.

HOW TO GROW
Plant in spring with a generous handful of garden compost incorporated into the planting area. Keep moist through spring and early summer, but do not allow to become waterlogged. Propagate by division after flowering in spring (see Dividing an herbaceous perennial alpine, page 54) or by seed (see below).

GROWING TIP
Sow seed while it is still green, as soon as it is dispersing in summer (see Growing alpines from seed, page 60).

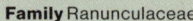

Family Ranunculaceae	
Height & spread 5–10x10–15cm/ 2–4x4–6in	
Flowering time Spring	
Hardiness H6	
Position Part shade	

A LEAF FOR THE LIVER
Liverleaf gets its common name from the folkloric use of it to treat medical conditions of the liver in Europe.

Bukhara iris

Iris bucharica aka corn-leaf iris

This ephemeral beauty is from the mountains of central Asia. It has thin, elegant, fresh green foliage topped with yellow/white blooms, to 40cm/16in tall, in early spring.

—

WHERE TO GROW
Plant outdoors in full sun in well-drained soil or in a pot, with protection from excessive rain.

HOW TO GROW
Add plenty of grit to the planting site if outdoors or grow in a more than 50 per cent inorganic mix (e.g., 30 per cent John Innes No. 3 and 70 per cent grit/pumice) in a container. Ensure the soil is moist through the growing season but keep it dry as soon as leaves start to die back. Bukhara iris can also be grown in pots in a sand plunge (see Winter protection, alpine houses and sand plunges, page 22) Propagate by carefully lifting and dividing plants in late summer (see Division, page 40) or by seed sown in summer (see Growing alpines from seed, page 60) – germination can be erratic.

GROWING TIP
The tuberous roots attached to the bulb are brittle, so it is best to buy pot-grown Bukhara iris to plant in spring rather than as bulbs of a dormant plant.

Family Iridaceae	
Height & spread 20–40x15–30cm/ 8–16x6–12in	
Flowering time Early spring	
Hardiness H3–H4	
Position Full sun	

THE JUNO IRIS
The Bukhara iris belongs to a section in its genus known as 'Juno irises' (*Iris* subg. *Scorpiris*). The alpine department of the Royal Botanic Gardens, Kew holds the national collection of these irises, and looks after hundreds of them.

Pygmy iris

Iris pumila aka dwarf bearded iris

This wonderfully petite bearded iris species comes in a variety of flower colours from purple to yellow to white. Its leaves reach only about 10cm/4in long, and its flowers are held just above.

—

WHERE TO GROW

Plant in a rockery, raised bed or trough in well-drained, gritty soil. Ensure pygmy iris is positioned in full sun, and is not overshadowed, to encourage flowering.

HOW TO GROW

Set its ginger-like rhizome just at or below the soil surface when planting in spring. Feed monthly with balanced fertilizer in the growing season. Propagate after flowering by separating rhizomes with growth points (see Dividing an herbaceous perennial alpine, page 54).

GROWING TIP

Watch out for viruses (see page 136); these create distorted growth and irregular streaking on the foliage. Infected plants must be destroyed.

Family Iridaceae

Height & spread
5–10×10–20cm/
2–4×4–8in

Flowering time
Early summer

Hardiness H4

Position Full sun

OF HYBRID ORIGIN
Iris pumila is a naturally occurring hybrid caused by the crossbreeding of two species (*I. attica* and *I. pseudopumila*) where they co-occur in the wild.

Reticulate iris

Iris reticulata aka golden netted iris

A rock garden favourite is this species of central Asian iris, which is available in a wide range of flower colours. Its grassy leaves form in winter/early spring, as do the blooms.

—

WHERE TO GROW
Grow in full sun in well-drained soil with plenty of added grit in a rock garden, raised bed or pot. The bulbs enjoy being dry and warm for their dormancy, so plant where they will dry out in summer.

HOW TO GROW
In autumn, plant reticulate irises at least 5cm/2in deep in an open position where they will not get overshadowed. Lift and divide the bulb clumps in autumn, to spread to different locations (see Division, page 40).

GROWING TIP
Bulbs can diminish over time, so feed with liquid fertilizer every two weeks in the growing season, to keep them strong.

Family	Iridaceae
Height & spread	10–15x5–10cm/ 4–6x2–4in
Flowering time	Winter–spring
Hardiness	H7
Position	Full sun

A GOLDEN NET
The reticulate (aka golden netted) iris is so named because of its papery netted casing around each bulb.

NOTABLE CULTIVARS
- 'George' has purple flowers with yellow marks on the dark lower petals.
- 'Katharine Hodgkin' is an old-fashioned favourite, with pale blue flowers marked with yellow and with deep blue veins; it is long-lived in a garden.
- 'Painted Lady' is a very pale form with cream and streaked, light blue petals.

Alpine juniper

Juniperus communis var. *saxatilis* aka *J.c.* subsp. *alpina,*
J.c. subsp. *nana*

This dwarf and prostrate evergreen shrub will keep your
rock garden green all year. Its scented prickly foliage creeps
slowly over the ground, and black fruits form in autumn.

—

WHERE TO GROW
Alpine juniper is suitable for a range of cultural conditions,
but prefers well-drained soil with an open aspect, for
example, a rock garden or raised bed.

HOW TO GROW
Plant in spring and water in well, to give the whole season
for establishment. Although drought-tolerant once
established, water through dry spells in the first few years
after planting. Propagate by semi-ripe cuttings in autumn
(see page 36) or by layering in spring (see page 39).

GROWING TIP
Alpine juniper takes well to light pruning, but this is
required only if its growth is covering other plants.

Family Cupressaceae

Height & spread
10–50x30–100cm/
4–20x12–39in

Flowering time
Spring

Hardiness H7

Position Full sun

A TRUE PLANT OF THE WORLD
Juniper (*J. communis*) has the
largest distribution of all woody
plants. It grows from the edges
of the Arctic Circle down to
Mediterranean, North Africa
and across the whole northern
hemisphere. The alpine juniper
(*J.c.* var. *saxatilis*) is the dwarf form
specially adapted to subarctic
and alpine habitats.

Edelweiss

Leontopodium nivale aka cat's paws, cliffhanger flower, lion's paw, wool flower

A popular musical and film called *The Sound of Music* made this iconic alpine plant globally famous. Its small rosettes of silvery foliage are topped by hairy, white, star-shaped flowers, 15cm/6in high.

—

WHERE TO GROW

Edelweiss is perfect for a rock garden or raised bed with good drainage. Being a true alpine, it likes cool and bright conditions.

HOW TO GROW

Plant in spring in gritty soil, somewhere with no competition from other plants. Propagate from seeds; collect them when the flower head starts to dry and sow the seed fresh (see Growing alpines from seed, page 60).

GROWING TIP

Position edelweiss where it gets full sun for at least a few hours daily, avoiding hot afternoon sun.

Family Asteraceae

Height & spread
15–20x10cm/6–8x4in

Flowering time
Spring

Hardiness H7

Position Full sun–part shade

A SYMBOL OF LOVE
Because this strange alpine flower grows in such hard-to-reach places, it has traditionally been given as a symbol of dedication to a lover.

Planting a rock garden

A rock garden is the ultimate place to show off your alpine and rockery plants. However, the choice of plants for a rock garden and how they are added can make a big difference to its success.

When working out what plants you want in your rock garden, make sure you utilize all the various tiers and planting pockets. You should consider buying three of each species and group these together in the rock garden to create instantly larger, naturalistic clumps. Also, make sure you thoroughly research the ultimate spread of each species, so you don't overplant. It's better to have fewer, well-grown plants than too many all competing for space. Avoid buying too many different plants, too.

A planting scheme can look very flat if all the plants are the same size and texture, so you will need a few larger specimen plants and a small shrub or tree to add height. Don't forget to include some bulbs, too (see Planting a patch of early squill bulbs, page 118). See also Choosing and buying rock garden plants, page 26.

You can plant at any time of year, but spring is ideal as it allows for the longest time to establish. Avoid planting if the ground is sodden or frozen, or if conditions are very hot and dry.

Make sure that you water all the pots at least a few hours before planting, so the plants are hydrated and have had time to drain. After planting and topdressing, water the whole area thoroughly. You can always check how moist the soil is by scratching away the topdressing grit and touching the soil below.

It is a good idea to water a rock garden through its first summer after planting, and also in really dry spells you need to do this at least every few weeks.

1. Place and plant your largest plants first (here, *Agave havardiana* and crimson bromeliad/*Fascicularia bicolor*).
2. Arrange the small pots (here including *Azorella*, *Lewisia*, *Leontopodium* and *Armeria*) across the space, grouping species together to look more natural. Bear in mind the spread of each when mature, so they will not crowd each other by then. Lay the bulbs (here, *Scilla* and *Crocus*) out at this time.
3. Dig a planting hole. Then remove a plant from its container and carefully loosen the root ball, to tease out the roots and encourage growth. Insert the plant in the hole so the soil is at the same height as when the plant was in its pot or plant slightly proud of the soil.
4. Use vertical cracks to plant at an angle; don't be afraid to squash the root ball lightly to fit it in the hole.
5. Tuck grit topdressing around the neck of each plant so it doesn't touch the soil. Then cover the surface between plants with 3cm/1¼in-thick layer of grit, to suppress weed growth.

Siskiyou lewisia

Lewisia cotyledon aka cliff maids

The mountains of California and Oregon are home to this succulent. It produces a thick taproot, a rosette of rubbery leaves and multi-stemmed clusters of stunningly bright flowers, 15cm/6in high, in assorted colours.

—

WHERE TO GROW

Grow in a well-drained rock garden, trough or pot. Siskiyou lewisia requires a bright position, where it won't be overshadowed by surrounding plants.

HOW TO GROW

Plant in spring at an angle, to increase drainage. Minimal water is required. Propagate by seed in spring (see Growing alpines from seed, page 60). Alternatively, in spring, cut off a rosette with a 5cm/2in stem, using a sharp knife; allow the cutting to dry for a week before planting it in a well-drained potting mix (see Succulent stem cuttings, page 38).

GROWING TIP

Siskiyou lewisia is suitable for planting in a north- or east-facing drystone wall.

Family Montiaceae
Height & spread 10x15cm/ 4x6in
Flowering time Late spring–summer
Hardiness H4
Position Full sun–part shade

HOLDING ON TO BREATH

Lewisia uses crassulacean acid metabolism (CAM) photosynthesis, which is a technical term for saying that they breathe only at night. This enables them to keep their pores closed in the day, thereby preventing water loss in dry environments.

Maihuén

Maihuenia poeppigii

This extremely prickly, cushion-forming cactus hails from the high mountains of Chile and Argentina. Its cylindrical stems bear tiny succulent leaves, long spines and pale yellow flowers held among the cushion.

—

WHERE TO GROW
Maihuén is safest planted in a trough or pot, which can be moved indoors in winter. However, it can be grown outside with some protection from winter wet (see Winter protection, alpine houses and sand plunges, page 22). Provide lots of light and good ventilation.

HOW TO GROW
Plant in sharply drained potting mix. Water monthly with low-nitrogen fertilizer in summer. Propagate by stem cuttings, which are easy to root (see Succulent stem cuttings, page 38).

GROWING TIP
Shelter from summer heat by positioning maihuén away from afternoon sun; it is a high-altitude plant.

Family Cactaceae	
Height & spread 10–15x50cm/4–6x20in	
Flowering time Summer	
Hardiness H5	
Position Full sun–part shade	

WHAT'S IN A NAME
This species is named after the German botanist and explorer Eduard Pöppig, who travelled all over South America between 1827 and 1832.

Grape hyacinth

Muscari neglectum aka common grape hyacinth

Grape hyacinth is native to the Mediterranean basin and eastwards to central Asia. Spikes of purple flowers with white tips form in spring on stems to 15cm/6in tall, over grassy foliage.

—

WHERE TO GROW

This bulb is easy to grow in a container as well as in a rock garden or raised bed. It prefers a well-drained, sunny position, to flower well. Its long foliage grows from autumn to summer, so position grape hyacinth away from smaller plants so they are not overshadowed.

HOW TO GROW

Grow in a mix of equal parts grit and John Innes No. 3 in a container; or add a handful of grit to the planting hole in the garden. Propagate by lifting and separating bulb clumps in late summer (see Division, page 40).

GROWING TIP

Grape hyacinth can become invasive if left unchecked. Therefore, remove the seed heads after flowering, to slow it down.

Family Asparagaceae	
Height & spread 5–15x5–30cm/ 2–6x2–12in	
Flowering time Spring	
Hardiness H6	
Position Full sun	

A GRADUATION OF COLOUR

The inflorescence of grape hyacinth is made up of many individual flowers, which start opening at the bottom and work their way to the top, changing from a light to a dark colour as they go.

Dwarf narcissus

Narcissus spp. aka dwarf daffodil

This group of miniature daffodils comes mainly from the Mediterranean region. All species of dwarf narcissus have a central trumpet (corona) and six outer petals (tepals), but they vary in scent, shape, colour and size – although most are around 10cm/4in tall.

—

WHERE TO GROW
Position in moist but well-drained soil in an open aspect, in a rock garden or raised bed. A container also works well.

HOW TO GROW
Plant dormant bulbs in autumn, adding handfuls of garden compost and grit to the planting site. Position at twice the depth of each bulb below the soil surface. Keep moist through the growing season and feed with balanced liquid fertilizer every two weeks. Propagate by lifting and dividing bulb clumps in autumn (see Division, page 40).

GROWING TIP
Watch out for diminishing blooms, which could be a sign of virus (see Pests and diseases, page 135) or narcissus bulb fly (page 135).

Family Amaryllidaceae	
Height & spread 10–30x10–20cm/ 4–12x4–8in	
Flowering time Spring	
Hardiness H5–H6	
Position Full sun	

Narcissus cyclamineus

NOTABLE SPECIES

- *N. bulbocodium* (hoop petticoat daffodil) bears bright yellow flowers each with a wide-fluted corona and small outer tepals.
- *N. cuneiflorus* (aka *N. asturiensis*; pygmy daffodil) is a perfectly proportioned, miniature, yellow daffodil, 10cm/4in tall.
- *N. cyclamineus* (cyclamen-flowered daffodil) has a long corona and reflexed outer tepals in yellow.
- *N. jonquilla* (jonquil) is one of the most highly scented species; it produces multiple blooms per stem, each with a reduced corona and large tepals.
- *N. triandrus* (angel's tears) varies in floral colour between white and yellow; its blooms have a delicately pendulous habit.

NARCISSISTIC NARCISSUS
The genus *Narcissus* takes its name from the Ancient Greek mythological character of the same name. He sat at the edge of a pool of water unable to break away from his reflection. In some versions of the story, he transformed into the narcissus flower, and still faces downwards towards the water. The word to describe a self-centred individual (narcissist) is also from this source.

Little prickly pear

Opuntia fragilis

Challenging your assumptions about cacti, this hardy little prickly pear comes from the mountains of North America. The paddles (aka thickened stems) grow slowly, and it has bright yellow flowers, 5cm/2in high.

—

WHERE TO GROW

Grow in sharply drained soil in a rock garden, crevice garden, pot or trough. Little prickly pear requires lots of sun to flower well.

HOW TO GROW

Plant in spring, to give this cactus a long growing season. Propagate by stem cuttings in spring (see Succulent stem cuttings, page 38); use barbecue tongs to handle the paddles and avoid the spines.

GROWING TIP

Take care when gardening near this plant, because its paddles are easily detached by snagging on clothing.

Family Cactaceae	
Height & spread 10x30cm/4x12in	
Flowering time Summer	
Hardiness H7	
Position Full sun	

TOUGH AS OLD BOOTS
Little prickly pear is the most northerly growing cactus species, happily thriving in the wild as far north as the middle of Canada.

Moss phlox

Phlox subulata aka creeping phlox, moss pink

Moss phlox is a beautiful, evergreen, mat-forming alpine from the mountains of North America. Starry flowers in pink, purple or white smother the surface of the needle-like foliage in spring and early summer.

—

WHERE TO GROW

Grow at the front of a raised bed or rock garden in well-drained soil in full sun. Moss phlox is also good for a trough or a larger pot.

HOW TO GROW

Plant in spring, adding grit to heavier soils. Keep moist through the first season of growth and through dry spells. Trim back around the edges if moss phlox encroaches on other plants. Propagate by softwood cuttings in summer (see page 36).

GROWING TIP

Different varieties of moss phlox can be planted near each other, for a riot of colour during late spring/ early summer.

Family Polemoniaceae	
Height & spread 5–10x10–50cm/ 2–4x4–20in	
Flowering time Spring–early summer	
Hardiness H6	
Position Full sun	

PINK MOON

The first full moon in April is named the 'pink moon' in *The Old Farmer's Almanac*, a traditional North American periodical publication. This is not because the moon is pink, but because it coincides with the flowering of moss pink, with its (often) bright pink flowers.

Casting a hypertufa trough

Stone troughs are the classic way to grow and display small alpine plants. Historically, these have been old stone drinking troughs or sinks. If you don't have the budget to splash out on an authentic stone one, here is a simple and cost-effective way to create a realistic imitation. It uses hypertufa, which is a lightweight artificial stone mix that mimics tufa, and is easy to cast into any shape.

This hypertufa trough is made from a mixture of cement, sand, perlite and sieved compost in equal proportions. However, you can experiment and change the proportions of these to make it more like different stone, but never decrease the amount of cement to less than a quarter of the total volume or you may get a weak set. Cement and perlite produce fine dust, so always wear gloves, a dust mask and eye protection throughout the whole mixing process.

You need to build a mould to cast your trough in: this can be of wood (as here), polystyrene or reinforced cardboard. Line the mould with plastic so the hypertufa mix doesn't stick. You also need to be able to dismantle the mould from the outside once the trough is set, so just screw it together. The chicken wire sandwiched in the base reinforces the structure; remember to cut holes in the wire where the drainage holes will be.

The trough needs to be placed in a frost-free environment while the cement cures. Cement curing is a chemical reaction with water, and keeping the hypertufa mix cool and humid creates a stronger cure that will be more frost-resistant. It should set hard within 12–24 hours, while the main cure takes about a week. The trough should be wrapped with plastic the whole time. After this, it can be uncovered and left in the rain to wash away alkalinity before planting.

Your hypertufa trough will look very dark in colour initially but will lighten with age, and eventually lichens and mosses will grow on it.

YOU WILL NEED

- Sawn timber, at least 2.5cm/1in thick (depending on the size of trough being built)
- Saw
- Electric drill and drill bits
- Wood screws and screwdriver
- Marine ply (for base of mould)
- Plastic sheeting or similar (to line and cover the mould)
- Masking tape
- 2 cardboard tubes, 6cm/2½in long
- Chicken wire (for strengthening base of trough) and wire cutters
- Portland cement
- Builders' sharp sand
- Perlite
- Potting compost, sieved
- Thick gloves
- Dust mask
- Eye protection
- Large bucket
- Trowel (for mixing ingredients)
- Large watering can or bucket
- Timber offcut (for a ramming tool)

1. Build the inner and outer frames for the mould by cutting the timber into eight pieces of appropriate sizes, four for each frame. Then drill and screw the sides together, the inner frame must be able to be dismantled from the inside. Place the outer frame on the marine ply board. Line both inner and outer frames with the plastic sheeting and secure in place with masking tape. Tape the cardboard tubes in the middle of the outer frame base, as drainage holes. Cut a piece of chicken wire to slightly smaller than the base and include holes for the cardboard tubes; set to one side.

2. Measure your dry ingredients and mix in a bucket. Add water slowly and stir in well until you get the texture of thick cottage cheese.

3. Pack the base with a 3cm/1¼in layer of hypertufa mix, position the prepared piece of chicken wire and then add a further 3cm/1¼in layer of mix over it.

4. Position the inner frame inside the outer one. Then add mix down the sides, compacting as you go.

5. Cover the mould with a large plastic sheet and tape in place, to keep the hypertufa mix moist while it starts to cure.

6. After 12–24 hours, carefully remove the mould, and add texture to the surface with the trowel. Re-cover with the plastic sheet for at least another week.

Tufted horned rampion

Physoplexis comosa aka *Phyteuma comosum*

Tufted horned rampion is a remarkable European alpine producing clusters of bottle-shaped, pale lavender flowers with tapered, dark purple points, on stems 7cm/3in tall. It has a rosette of serrate, heart-shaped leaves.

—

WHERE TO GROW
Grow in part shade in an alpine house in a pot or trough or in a sheltered, raised, well-drained position in a rock garden.

HOW TO GROW
Being native to limestone cliffs in the European Alps, it grows well planted on tufa (see Planting into tufa, page 112). In a container, grow in a mix of at least 50 per cent inorganic material such as pumice or grit and the rest John Innes No. 3. Keep from drying out in the growing season and do not overwater in winter. Tufted horned rampion can be prone to slug damage (see Pests and diseases, page 134).

GROWING TIP
The seed of this herbaceous perennial is tiny, so crush whole capsules to release it. Sow in autumn with the debris in order to not lose any seeds (see Growing alpines from seed, page 60).

Family	Campanulaceae
Height & spread	5–10x5–10cm/ 2–4x2–4in
Flowering time	Summer
Hardiness	H6
Position	Part shade

A LOCAL CURIOSITY
This botanical oddity is a localized endemic species, occupying just 500 sq. km/200 sq. miles of limestone cliffs in the north-eastern Italian Alps – its range just extending across the border into Austria and Slovenia.

Auricula

Primula auricula aka mountain cowslip

Auricula is a showcase alpine plant, native to the mountains of central Europe, and it has been selectively bred by collectors to have flowers, 15cm/6in high, in every colour of the rainbow.

—

WHERE TO GROW

This primrose relative will happily grow in a well-drained rock garden, pot or trough positioned out of hot sun.

HOW TO GROW

Plant in spring, and feed with dilute fertilizer every spring, especially if growing in a pot. Keep auricula clear of debris and competition from surrounding plants; it will then grow happily without disturbance for several years. Propagate by basal cuttings/divisions in spring (see page 37).

GROWING TIP

Keep an eye out for pests, because auricula's thick roots and stems are tasty to vine weevil larvae (see page 136).

Family Primulaceae	
Height & spread 10–20x10cm/4–8x4in	
Flowering time Spring	
Hardiness H5	
Position Part shade	

GROWING IN TECHNICOLOUR

The craze for auricula cultivation has driven the development of thousands of *Primula* varieties and even bespoke 'auricula theatres' to display them in.

Shooting star

Primula meadia aka *Dodecatheon meadia*, American cowslip

Shooting star – a summer-dormant herbaceous perennial – grows in the forests and prairies of eastern North America. It develops a rosette of simple leaves in spring, soon followed by clusters of pink flowers, with reflexed petals and multicoloured centres, on stems up to 50cm/20in high.

—

WHERE TO GROW

Grow in soil with added garden compost in a relatively open position in a rock garden or raised bed that is moist in winter and drier in summer. Shooting star can also be grown in a pot, but it is then harder to manage its dormancy.

HOW TO GROW

Plant in spring in organic-rich soil, adding a handful of garden compost when planting. Propagate by division in early spring (see Dividing an herbaceous perennial alpine, page 54) or by seeds sown fresh in summer (see Growing alpines from seed, page 60).

GROWING TIP

The blooms of shooting star also come in white and paler shades of pink; they contrast nicely when planted together.

Family	Primulaceae
Height & spread	20–50x15–20cm/ 8–20x6–8in
Flowering time	Spring–early summer
Hardiness	H5
Position	Full sun–part shade

UPSIDE DOWN PRIMROSE
Shooting star is very closely related to the common primrose (*P. vulgaris*), native to Europe. However, its flowers face downwards, and its petals are bent backwards. This makes them resemble a badminton shuttlecock in flight – hence the common name 'shooting star'.

Mountain cherry

Prunus prostrata aka creeping cherry, rock cherry

This miniature deciduous cherry grows high in the
mountains around the Mediterranean and in central Asia.
Its spreading branches are smothered with the classic,
pale pink cherry blossom in spring.

—

WHERE TO GROW

Grow in well-drained soil in a sunny sheltered position
in a raised bed or rock garden such as a sunny rock
crevice (see Crevice gardens, page 18).

HOW TO GROW

Plant in spring, ensuring the soil is well-drained
by adding half a bucket of grit to the planting
area. Keep watered during its first year through
dry spells; mountain cherry is drought-tolerant
once established. Propagate by hardwood cuttings
during winter (see page 36) or by softwood
cuttings in early summer (see page 36).

GROWING TIP

The flowers may get frosted in very cold climates, below
−5°C/23°F, in mid-spring. To keep frost off the blooms,
mountain cherry may need protection in spring, or plant
it near a wall.

Family Rosaceae	
Height & spread 15–100x30–100cm/ 6–39x12–39in	
Flowering time Spring	
Hardiness H3–H4	
Position Full sun	

A WILD CROP RELATIVE

The mountain cherry is closely
related to our domesticated
Prunus varieties such as plums
and cherries. There is now
increased interest and research
into these crops' wild relatives
and into genetics conferring
disease or drought-tolerance.

Pasqueflower

Pulsatilla vulgaris

This European wild flower is endangered in the UK but is popular in cultivation in temperate regions of the world. Hairy purple flowers, 30cm/12in high, are followed by a mound of woolly dissected foliage and fluffy seed heads.

—

WHERE TO GROW

Grow in sunny, evenly moist soil in a rock garden or raised bed. Pasqueflower does not enjoy sitting in wet conditions, and it resents any movement of its long, easily damaged, woody taproot.

HOW TO GROW

Pasqueflower leaves can grow quite large, so allow at least 30×30cm/12×12in between plants. Plant in spring or autumn, to establish well. Propagate by seeds in spring (see Growing alpines from seed, page 60).

GROWING TIP

Make sure you carefully cut back the old foliage in early spring, to show off the flowers, which emerge before the new-season leaves.

Family Ranunculaceae

Height & spread
30x 30cm/12x12in

Flowering time
Spring

Hardiness H5

Position Full sun

A FLOWER OF SPRING

The common name pasqueflower derives from the Latin *paschalis* (relating to Easter) from the Hebrew *Pesach* (Passover), which also gave rise to the modern French *Pâques* (Easter). Thus, these spring festivals are associated with the flowering time of pasqueflower.

Vegetable sheep

Raoulia

Vegetable sheep is a genus of mat- and cushion-forming alpines from New Zealand. The tiny hairy leaves grow very tightly, producing a very compact habit. Vegetable sheep generally produces reduced aster-like flowers.

—

WHERE TO GROW

Some species can be grown outside in a well-drained rock garden or trough, but most are best in a pot in a cool alpine-house sand plunge (see Using a sand plunge for alpine plant cultivation, page 128).

HOW TO GROW

Provide good light and ventilation; shade from hot sun. Grow in a mix of garden compost (40 per cent) and inorganic drainage material such as grit or pumice (60 per cent). Keep moist in the growing season but withhold water in winter. Feed minimally, to encourage compact growth and to discourage pests. Propagate by cuttings in spring (see Propagating a cushion-forming plant, page 66).

GROWING TIP

To prevent the hairy foliage from wetting and rotting, try not to water these plants from overhead.

Raoulia albosericea

Family Asteraceae
Height & spread 1–10x10–50cm/ ½–4x4–20in
Flowering time Summer
Hardiness H4–H5
Position Full sun–part shade

A EWE-NIQUE PLANT

In the landscape of the New Zealand Alps, true vegetable sheep (*R. eximia*) can form large 'flocks' of cushions, each literally the size of sheep. Their dense 'woolly' growth further adds to the mimicry.

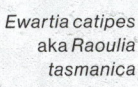

Ewartia catipes aka Raoulia tasmanica

NOTABLE SPECIES

*   *R. australis* (scabweed), one of the more well-known species in cultivation, has silvery foliage and yellow flowers; grows best in an alpine house or a sheltered trough.
*   *R. eximia* (New Zealand vegetable sheep, true vegetable sheep) forms exceptionally dense cushions of pale blue/white foliage; is suitable for alpine-house cultivation.
*   *R. tenuicaulis* (mat daisy, tutahuna) is a green-leaved, mat-forming species, very closely hugging the ground; it grows well in a well-drained rock garden bed.

Alpenrose

Rhododendron ferrugineum aka snow rose

The alpenrose is an iconic plant of the European Alps, painting the mountainsides pink when in bloom. It's a small evergreen shrub with dark glossy green leaves and clusters of garish trumpet flowers.

—

WHERE TO GROW

Grow in an open position in acidic, organic-rich soil in a rock garden or raised bed. Ensure the area you plant it in does not get waterlogged in winter.

HOW TO GROW

Plant in spring after the last frosts, adding ericaceous compost to the planting hole. Prune out only dead or diseased stems after flowering. Propagate by semi-ripe cuttings in late summer (see page 36) or by layering in spring (see page 39).

GROWING TIP

Rhododendrons can be susceptible to vine weevil damage – watch out for the distinctive notches on the leaves left by the adults (see page 135).

Family Ericaceae

Height & spread
30–100x50–100cm/
12–39x20–39in

Flowering time
Summer

Hardiness H6

Position Full sun

A HAIRY SIBLING
The closely related species, the hairy alpenrose (*R. hirsutum*), also grows in the European Alps. Unlike *R. ferrugineum*, it prefers alkaline, lime-rich soil. The two soils (and the two species) frequently co-occur in the wild.

Himalayan ginger

Roscoea purpurea aka hardy ginger

Himalayan ginger will bring a little piece of the East to your garden. It blooms on stems 30cm/12in above the ground, producing stunning, orchid-like flowers in purple in late summer.

—

WHERE TO GROW

Being a plant of the mountain slopes of the Himalayas, it grows best in sheltered conditions, liking moist, organic-rich soil in part shade. Position in a rock garden or raised bed.

HOW TO GROW

Plant in the ground where you can provide ample moisture in the growing season, and add plenty of organic material such as garden compost to the planting hole. In climates where temperatures regularly dip below −15°C/°5F Himalayan ginger may not be hardy, so apply a mulch, 5cm/2in thick, over the plant in winter.

GROWING TIP

Propagate by division in early spring (see page 40) or by seed sown in late autumn (see pages 40).

Family Zingiberaceae	
Height & spread 20–30x15–40cm/ 8–12x6–16in	
Flowering time Late summer	
Hardiness H5	
Position Part shade	

NOTABLE CULTIVAR AND OTHER NOTABLE SPECIES

- 'Red Gurkha' is the red-flowered form of *R. purpurea*; it often has red stems, too.
- *R. alpina* grows up to 3,600m/11,800ft above sea level; its pale purple blooms appear before the leaves fully emerge.
- *R. cautleyoides* bears pale yellow flowers and is one of the first Himalayan gingers to bloom, in early summer.
- *R. humeana* bears purple flowers, which emerge earlier than *R. purpurea* by a few weeks; like *R. alpina*, the blooms form before the leaves have fully emerged.
- *R. scillifolia* has long thin leaves and much smaller flowers, 3cm/1¼in wide.

A DIFFERENT KIND OF GINGER

Himalayan ginger is in the same family (Zingiberaceae) as the ginger that we cultivate for its tasty rhizome (*Zingiber officinale*). However, Himalayan ginger (*Roscoea*) is cultivated for ornamental purposes. It has a much-reduced rhizome with large tuberous roots that it refreshes with each growing season.

Net-leaved willow

Salix reticulata aka snow willow

Dwarf willows, such as the net-leaved willow, are a regular sight in alpine environments. This species has stunning, dark green, deciduous leaves with deeply cut veins over their surfaces. Little upright catkins, 5cm/2in long, form in spring.

—

WHERE TO GROW

Grow in full sun in a damp position and cool climate where the temperature does not regularly go above 30°C/86°F. Choose a moist but well-drained position in a rock garden or raised bed where you can get close and appreciate the foliage and flowers.

HOW TO GROW

Before planting, add plenty of organic material to the planting area. Propagate by hardwood cuttings in winter (see page 36) or by layering in spring (see page 39).

GROWING TIP

In an area where temperatures may soar above 30°C/86°F, you may have to consider planting in a partially shaded, cool position, to shelter net-leaved willow from hot summer sun.

Family Salicaceae

Height & spread
5–10x10–30cm/
2–4x4–12in

Flowering time
Spring

Hardiness H7

Position Full sun–part shade

A TRUE ARCTIC-ALPINE 'TREE'
The net-leaved willow has an arctic-alpine 'circumpolar' distribution. This means that it grows in the north edge of all the continents south of the North Pole – North America, Europe and Asia. It is highly adapted to its freezing, wind-blown life by having branches that hug the ground; some are even subterranean.

Encrusted saxifrage

Saxifraga sect. Ligulatae and *S.* sect. Porphyrion aka rockfoils

These two groups of cushion-forming saxifrages are mainly native to Europe. Their leaves are encrusted with calcium carbonate excreted from hydathode glands. Spikes of starry flowers, up to 30cm/12in high, form above the foliage.

—

WHERE TO GROW
Grow in an open, north- or east-facing position that is shaded from the hottest part of the day, in a rockery or raised bed, with at least 50 per cent added grit.

HOW TO GROW
Plant in autumn or spring, to establish well. Individual rosettes die after flowering so deadhead them (see Pruning, page 32). Propagate by cuttings during early spring (see Propagating a cushion-forming plant, page 66).

GROWING TIP
Try growing these calciphiles (plants that love growing in limestone) in tufa (see Planting into tufa, page 112).

Family Saxifragaceae	
Height & spread 5–15x10–30cm/ 2–6x4–12in	
Flowering time Early summer	
Hardiness H7	
Position Full sun–part shade	

MINERALS FROM SPACE
Studies at Cambridge University in 2018 examined the calcium carbonate in saxifrage leaves and discovered that it was in a rare form called vaterite. This is unusual on planet Earth, and can have been sourced only from meteorites and other objects from outer space.

NOTABLE SPECIES AND CULTIVARS
- *S. cochlearis* is an easy-to-grow, cushion-forming species with spoon-shaped leaves and multi-stemmed flowers; 'Snowflake' bears white blooms.
- *S. cotyledon* 'Southside Seedling' is one of the largest encrusted saxifrages, spreading to form a mat with stunning, minutely red-speckled, white petals.
- *S. longifolia* 'Tumbling Waters' produces giant flower stems, to 50cm/20in, from large rosettes in a cushion.
- *S. oppositifolia* is a small prostrate species with pink/ purple flowers; it prefers cool temperatures averaging below 20°C/68°F.
- *S. paniculata* is an easy-to-grow, mat-forming species with white flower spikes to 30cm/12in tall.
- *S. squarrosa* (Dolomites saxifrage) is one of the smallest saxifrage species; it forms a tiny cushion with leaves only 2mm/^{1}/12in long.

Saxifraga cotyledon

Planting into tufa

A planted tufa rock is perfect as the centrepiece for a large pot or trough, while smaller pieces work well in small pots. Plants most suitable for tufa cultivation are naturally slow-growing and lime- (alkaline-) lovers, including many cushion-forming plants. Saxifrage (*Saxifraga*), *Dionysia*, houseleek (*Sempervivum*) and stonecrop (*Sedum*) are just a few alpines to get you started.

Plant in early spring, when the growing season is just starting, and the conditions are not too warm and sunny. Use very young plants or cuttings as they will have small root systems, which are easier to handle. Tufa is very soft to drill, so usually doesn't require much effort, but pre-soaking it in water reduces the amount of dust when drilled. The plant roots will also be inserted into a wet environment.

The natural tufa shown here was acquired second-hand for use in the display in the Alpine House at the Royal Botanic Gardens, Kew (see also Rocks, rocks, rocks, page 24). It is already covered in natural crevices. Taking advantage of these not only makes the planting look more naturalistic, but it also helps the plants establish better.

You can also use the steps shown here to plant into hypertufa rocks, which you can make yourself (see Casting a hypertufa trough, page 100). However, the surface may be harder to drill unless the hypertufa is made with a high proportion (50 per cent or more) of compost.

Even if you do not want to take the more advanced step of tufa cultivation, you can still appreciate plants growing in tufa at botanic gardens and understand the process that the plants have been through to get there.

1 Wearing appropriate eye, ear and hand protection, use an electric drill fitted with a 1cm/½in-wide bit to make holes at least 10cm/4in deep holes in the tufa.
2 Prepare the plants (here, cobweb houseleek/ *Sempervivum arachnoideum*) by bare-rooting and washing off the soil. Water also helps keep the roots straight.
3 Feed the roots down into the rock, using a thin wooden skewer or chopstick.
4 Carefully pack around the roots with pure horticultural sand, crushed tufa or fine pumice.
5 Spray the plants carefully so you do not wash them out. Leave the planted tufa in a dish of water for a few weeks while the plants establish.

Mossy saxifrage

Saxifraga sect. Saxifraga aka rockfoils

These evergreen cushion plants are closely related to their encrusted cousins *Saxifraga* sect. Ligulatae and *S.* sect. Porphyrion (see page 111). However, mossy saxifrages are often from warmer climates and lower altitudes. Their dissected leaves are arranged in rosettes, and they produce five-petalled, white flowers, 30cm/12in high. They may develop gaps in the middle when mature – this is natural.

Family Saxifragaceae	
Height & spread 10–15x10–50cm/ 4–6x4–20in	
Flowering time Early summer	
Hardiness H5–H7	
Position Full sun–part shade	

WHERE TO GROW

Grow in a pot, raised bed, rock garden or trough in moist, well-drained soil. Although they tolerate more warmth than encrusted saxifrages, mossy saxifrages still prefer to be shaded from hot midday sun.

HOW TO GROW

Propagate by cuttings in spring (see Propagating a cushion-forming plant, page 66). Mossy saxifrages may seed themselves around, when happy.

GROWING TIP

Deadhead with scissors after flowering, to tidy up, or wait until the flower stems go brown and collect the tiny black seeds for sowing (see Growing alpines from seed, page 60).

Saxifraga rosacea

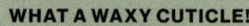

WHAT A WAXY CUTICLE!
Some of the mossy saxifrages such as *S. canaliculata* have a sticky coating of wax over their leaves that protects them from water loss.

Saxifraga rosacea

NOTABLE SPECIES AND HYBRID CULTIVARS

- *S. canaliculata* is a Spanish species that survives temperatures at or above 30˚C/86˚F for short periods and tolerates more sun than most other mossy saxifrages.
- *S. cebennensis* forms compact cushions and is more suitable for alpine-house growing.
- *S. hypnoides* enjoys wet conditions, so is perfect for areas with lots of rain and temperatures dipping down to –10˚C/14˚F.
- *S.* (Mossy Group) 'Peter Pan' is a vigorous, pink, free-flowering hybrid that makes a great garden plant.
- *S. rosacea* is a European native forming very large, prostrate mats with pale cream, starry flowers; it is similar to *S. hypnoides* in its cultivation requirements.
- *S.* 'White Pixie' is another vigorous, free-flowering cultivar with white flowers.

Strawberry begonia

Saxifraga stolonifera aka creeping saxifrage, mother of thousands

Often sold as a house plant, this hardy ground cover hails from temperate east Asia. Like a strawberry, it produces runners from the main rosette. Flower spikes, 35cm/14in high, develop in summer.

—

WHERE TO GROW
Strawberry begonia is great for a moist shady corner of a rock garden or for planting in a pot or trough.

HOW TO GROW
Add compost to the area before planting and water in well. After flowering, snip off the spent flower spikes. Propagate by runners in summer, which can easily be removed and potted up (see page 39).

GROWING TIP
This species has very shallow roots, so be careful when raking or weeding nearby not to accidentally uproot them.

Family Saxifragaceae	
Height & spread 15x50cm/6x20in	
Flowering time Early summer	
Hardiness H4	
Position Part shade–full shade	

HERBAL REMEDY
Modern studies have demonstrated strawberry begonia contains antioxidants, which have historically been used as a medicine to treat inflammation and infection.

London pride

Saxifraga × urbium aka none-so-pretty

London pride is a vigorous hybrid of Pyrenean saxifrage
(*S. umbrosa*) and St Patrick's cabbage (*S. spathularis*). Its
glossy dark green rosettes are enhanced by sprays of white
pink-speckled flowers, 30cm/12in high.

—

WHERE TO GROW
This is excellent ground cover for shaded parts of a rock
garden. It also tolerates dry conditions.

HOW TO GROW
Requiring minimum fuss, this is a true beginner's alpine.
Just tidy it up after flowering by removing spent flower spikes.
Propagate by division in spring (see Dividing an herbaceous
perennial alpine, page 54).

GROWING TIP
To get even more pleasure from London pride, why not
examine its flower petals in minute detail – they are full
of beautiful, multicoloured dots.

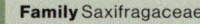

Family Saxifragaceae

Height & spread
10x50cm/4x20in

Flowering time
Early summer

Hardiness H5

Position Part shade–full
shade

PRIDE OF LONDON
Tradition states that this plant
spread from ruined gardens
over the rubble left behind by
the Second World War's London
Blitz, and that it came to represent
Londoners' strength and spirit.

Portuguese squill

Scilla peruviana aka Cuban lily, Peruvian lily

This stunning Mediterranean bulb is a must-have for warmer-climate rock gardens. The intricate inflorescences, formed of whorls of purple stars, perch 15cm/6in above the strappy green foliage.

—

WHERE TO GROW

Plant in a well-drained site in full sun; otherwise grow in a pot and move Portuguese squill under glass in cold weather.

HOW TO GROW

From spring to autumn, purchase actively growing potted plants and plant in soil with added grit, setting bulbs 5–10cm/2–4in below the surface. In a pot, use a 50:50 mix of John Innes No. 3 and grit. Fertilize with liquid feed every two weeks while actively growing. Propagate by lifting and carefully separating small side-bulbs in summer (see Division, page 40).

GROWING TIP

It's best to buy pot-grown plants of Portuguese squill (in active growth), rather than bulbs, as their roots often are easily damaged even when dormant.

Family Asparagaceae	
Height & spread 10–15x20–30cm/ 4–6x8–12in	
Flowering time Summer	
Hardiness H4	
Position Full sun	

PERUVIAN LILY OR PORTUGUESE SQUILL?
Even though Portuguese squill looks exotic, it is not from Peru as implied by its species name. The Peruvian name was given in 1753 using some notes from the late 1500s, and a mistake was made as to where the bulbs were from! Portuguese squill has kept its Peruvian species name ever since.

Planting a patch of early squill bulbs

Bulbs are a natural part of an alpine ecosystem and will help your rockery to feel naturalistic and wild. Most hardy bulbous species will happily thrive in the general rock garden soil without much additional treatment other than some granular fertilizer once a year, at the start of spring.

When positioning bulbs, it is important to consider how they fit in with the surrounding plants and their seasons of growth. For example, the squill (*Scilla*) species discussed here is a very early flowering bulb that requires lots of light. It will tolerate being planted at the base of an herbaceous plant that doesn't start to grow until late spring, but will not enjoy a position underneath an evergreen plant.

Planting bulbs in distinct groups allows you to better keep track of where they are and makes more focused displays of colour. To make them look more natural, try an irregularly shaped group, such as elongated triangle or oval.

There are a few practical benefits to planting along the edges of the rocks. Firstly, the rocks may act as a reminder when the bulbs are dormant, so you don't accidentally step on them or dig them up. Secondly, rocks can obscure the view of some of the bulbous leaves after they have flowered and are dying back.

Once the bulbs have finished flowering, you can cut off the spent blooms unless you intend to collect the seeds. Never remove the leaves until they go brown, as until then they will be feeding the bulbs for next year's growth.

1 In autumn, choose an unused area of rock garden and make sure it is weed-free. Then scrape back the grit topdressing.
2 Dig a flat-based hole, 10cm/4in deep and about 15cm/6in wide (here, for a handful of squill/*Scilla mischtschenkoana* bulbs).
3 Add a handful of garden compost to the planting area, then evenly space the bulbs in the hole with their growing points facing upwards.
4 Backfill the hole followed by a grit topdressing, 3cm/1¼in deep. Label the planted area.
5 Sit back and enjoy the floral display in spring.

Siberian squill

Scilla siberica

This dainty, blue-flowered bulb hails from the hills and woods of central Asia. Its gently nodding, star-shaped blooms grow in groups up to five on stems only 10cm/4in tall, above flat, bright green leaves.

—

WHERE TO GROW

Grow in moist but well-drained soil, ideally in a position that is sunny in winter, in a raised bed, rock garden, pot or trough.

HOW TO GROW

Plant dormant bulbs in autumn (see Planting a patch of early squill bulbs, page 118). Propagate by lifting and separating bulb clumps in autumn (see Division, page 40). Siberian squill will also self-seed freely in time.

GROWING TIP

Squill can become invasive if left unchecked, so deadhead after flowering to prevent seed spread if you do not want this.

Family Asparagaceae	
Height & spread 5–10x10–30cm/ 2–4x4–12in	
Flowering time Spring	
Hardiness H6	
Position Full sun–part shade	

A DIFFERENT KIND OF BLUEBELL

You would not be totally wrong for mistaking Siberian squill for an English bluebell (*Hyacinthoides non-scripta*). They are, in fact, closely related and in the same plant family, Asparagaceae.

OTHER NOTABLE SPECIES

- *S. bifolia*, as the name suggests, has two leaves per bulb; this European species produces darker blue blooms than *S. siberica* and red flower stalks.
- *S. mischtschenkoana* (Misczenko squill) is another central Asian species; it bears pale blue flowers and is squat and slower growing than the others described here.
- *S. verna* (spring squill) is a western European species, with very thin foliage and dense clusters of ephemeral, pale lavender flowers on the stem.

Stonecrop

Sedum

Among this large group of plants are short- and long-lived species. Commonly available hardy species are evergreen with a variety of foliage colours and jewel-like flowers, 15cm/6in high, in yellow, white and pink.

—

WHERE TO GROW

Stonecrop is useful for dry sunny areas of the garden. All this beginner's alpine needs is full sun and well-drained soil, so grow it anywhere it won't be overshadowed.

HOW TO GROW

Plant in spring or summer. Propagate by division in spring or summer (see Dividing an herbaceous perennial alpine, page 54), and replant with a small amount of grit over the stems.

GROWING TIP

Why not try planting a few distinct *Sedum* species near to each other, to bring out their differences?

Family Crassulaceae	
Height & spread 2–20x50cm/¾–8x20in	
Flowering time Summer	
Hardiness H6–H7	
Position Full sun	

Sedum acre

NOTABLE SPECIES, CULTIVARS AND SIMILAR GENERA

- *S. album* has cylindrical fleshy leaves and white flowers held in clusters above the plant.
- *S. dasyphyllum* spreads quickly and its individual blue leaves can root independently.
- *Petrosedum rupestre* (aka *Sedum reflexum*) has conifer-like foliage and grows to 15cm/6in tall. There is also a golden-leaved cultivar called 'Aureum'.
- *Phedimus kamtschaticus* (aka *Sedum kamtschaticum*) is a species from eastern Asia with foliage that dies back yearly and yellow flowers; it tolerates much wetter conditions than many other stonecrops; 'Variegatum' is a popular variegated form.
- *Phedimus spurius* (aka *Sedum spurium*) produces foliage and flowers in a variety of shades, from pink to almost red.

LIVING GOLD
The dense golden flowers on golden moss stonecrop (*S. acre*) in summer are a haven for bees on sea cliffs and rocks across Europe.

Repotting a houseleek

Houseleek (*Sempervivum*) is one of the easiest alpines to grow in a pot. The generic name *Sempervivum* is a compound of *semper* and *vivus* in Latin, translating as 'always-living' – alluding to its evergreen habit. Houseleek thrives on very little nutrition or water. After a few years, however, its pot may become weedy or it may start to rot if the drainage hole becomes blocked. A houseleek also bears 'monocarpic' rosettes, meaning each rosette dies once it has flowered. This can leave gaps in the display once the spent rosettes have been pulled out and discarded.

Repotting allows you to refresh the potting mix, increase the drainage and spread out the remaining rosettes to allow even growth and encourage flowering for the coming years. Discarding old compost also reduces the risk of overwintered soil pests such as vine weevil larvae (see page 136).

Plants purchased in garden centres may be congested or in poorly drained compost, so you may want to repot a newly bought houseleek to give it a good start in your care. This is best done in spring, when it is beginning to grow. Due to its thick succulent roots, houseleek is traditionally grown in a 50:50 mix of John Innes No. 3 and grit in a shallow terracotta pot.

After repotting and watering once, keep the pot somewhere bright and free of excess moisture for the next few weeks while the plant settles in, to avoid rotting (see page 134).

1. This pot of houseleeks (*Sempervivum minutum*), with its spent flowers and dead rosettes from last year, needs repotting. Remove the dead leaves and flowers.
2. Pull the living rosettes carefully free of the compost, using a dibber as necessary. Separate off the 'pups' (small rosettes).
3. Clean the pot and refill it with grit for the first 1cm/¾in, then add potting mix to just below the rim.
4. Use a dibber to make new holes for the rosettes. Insert the largest ones first, spreading them out. Then infill with the smaller rosettes.
5. Topdress with a thick layer of fresh grit around the rosettes and water in and label.

Houseleek

Sempervivum

These rosette-forming succulents are native to the mountainous regions of Europe, where they grow on thin soil in rock crevices and exposed dry mountain meadows. They have succulent, sometimes hairy, leaves and produce starry flowers on alien-like stems, to 15cm/6in high. Individual rosettes die after flowering. Growing a few assorted colour forms or leaf textures side by side enhances houseleek's differences.

Family Crassulaceae

Height & spread
5x20cm/2x8in

Flowering time
Summer

Hardiness H5–H7

Position Full sun

WHERE TO GROW

Plant in a rock garden, crevice garden (see page 18), pot or trough in full sun and with good drainage.

HOW TO GROW

To avoid rotting, plant in a 50:50 mix of John Innes No. 3 and potting grit in early spring, as houseleek starts to grow; then water once. When established, water every 1–2 weeks – even less in rainy weather. Raising the container with pot feet can help water to drain. Propagate by dividing offsets in spring (see Offsets/runners, page 39).

GROWING TIP

Traditionally, houseleeks are grown in shallow terracotta pots, because of their short succulent roots.

ANCIENT WISDOM
These alpines were traditionally planted on rooftops, not to plug water leaks, as their common name houseleek might suggest, but to ward off lighting strikes.

NOTABLE SPECIES AND CULTIVARS

- *S. arachnoideum* (cobweb houseleek) is covered with hairs.
- *S. arachnoideum* subsp. *tomentosum* is even hairier than the species; it prolifically produces small offsets, creating a mound of dense rosettes.
- *S. calcareum* 'Mrs Giuseppi' has large, blue-green rosettes with eye-catching, red tips.
- *S. globiferum* subsp. *allionii* (aka *Jovibarba allionii*) has very rounded rosettes with many 'pups' (offsets) that roll away from the parent plant.
- *S.* 'Red Beauty' has dark red foliage, which intensifies in sunny conditions.

Sempervivum tectorum

Blue-eyed grass

Sisyrinchium angustifolium aka narrow-leaved blue-eyed grass

A useful, summer-flowering rock garden plant is this
north-eastern American meadow plant, which has long,
grass-like foliage in tight fan clusters and bright purple,
six-petalled flowers, with yellow anthers, on stems to
25cm/10in tall.

WHERE TO GROW

Blue-eyed grass is ideal for well-drained soil in full sun
in a rock garden, raised bed, trough or pot.

HOW TO GROW

Plant in spring, adding grit to the planting site.
Blue-eyed grass is drought-tolerant once established.
Deadhead spent flower stems in autumn, and clear
out dead leaves in spring, to encourage new growth
and discourage rotting (see Rotting plants, page 134).
Propagate by division in spring (see page 40) or by seed
in autumn (see Growing alpines from seed, page 60).

GROWING TIP

A reliable, miniature, sterile hybrid cultivar that is
perfect for a rock garden or trough is *S.* 'E.K. Balls',
at only about 15cm/6in tall.

Family Iridaceae	
Height & spread 15–25x10–30cm/ 6–10x4–12in	
Flowering time Summer	
Hardiness H5	
Position Full sun	

INVASIVE POTENTIAL
Although blue-eyed grass is
native to North America, it has
been recorded in the wild across
the Mediterranean region in
France, Spain and Türkiye and as
far away as Japan.

Alpine snowbell

Soldanella alpina aka blue moonwort

The delicate alpine snowbell is iconic of the European mountains. Its 10cm/4in-tall racemes of pendulous, purple-bell flowers, with ragged edges, appear from early spring.

—

WHERE TO GROW

Grow in a moist position, at the front of a rock garden or raised bed, near a north-facing rock or wall to ensure alpine snowbell is cool. Shade it from hot summer sun, when dormant. Alpine snowbell is very easily outcompeted by more vigorous neighbouring plants, so keep it clear of competition.

HOW TO GROW

Plant in spring, adding a handful of garden compost to the planting area. Propagate by division after flowering, in spring (see Dividing an herbaceous perennial alpine, page 54).

GROWING TIP

Alpine snowbell can also be grown in a pot filled with a mix of one part garden compost, one part grit and one part John Innes No. 3, and placed in an alpine-house sand plunge (see Using a sand plunge for alpine plant cultivation, page 128).

Family Primulaceae	
Height & spread 5–10x10–15cm/ 2–4x4–6in	
Flowering time Spring	
Hardiness H5	
Position Full sun–part shade	

A SIGN OF SPRING

In the wild, this glorious little plant is a true harbinger of spring. It pops up and flowers following the receding snowline, drinking up the icy water from the melting snow.

Chilean blue crocus

Tecophilaea cyanocrocus

This half-hardy corm is well worth the effort to cultivate. Grassy foliage emerges in winter, quickly followed by six-petalled flowers, 10cm/4in tall. The unreal brightness of their blue colouring perhaps even surpasses the gentian (*Gentiana*; see trumpet gentian, page 80).

WHERE TO GROW

Grow in an alpine house in a pot in a sand plunge (see page 128). If temperatures in your area rarely drop below –5°C/23°F, plant Chilean blue crocus in a well-drained rock garden or raised bed.

HOW TO GROW

In autumn, plant corms 5cm/2in deep in a pot filled with a well-drained mix of at least 50 per cent inorganic drainage material (pumice/grit/clay granules) and the rest John Innes No. 3. Keep moist and frost-free through winter. While growing, feed every second week with liquid fertilizer; keep dry during summer, while dormant. Propagate by division of corms while dormant in autumn (see Division, page 40).

GROWING TIP

If attempting to grow Chilean blue crocus outside (it survives outside at the Royal Botanic Gardens, Kew), plant near a wall or large rocks, to provide warmth and shelter from summer rain.

Family	Tecophilaeaceae
Height & spread	10x10–20cm/4x4–8in
Flowering time	Spring
Hardiness	H3
Position	Full sun

A PLANT ON THE EDGE
This plant is highly threatened in its native Chile, because it has been over collected and grazed and its habitat destroyed. Chilean blue crocus was even thought extinct in the wild until its re-discovery in 2001.

Using a sand plunge for alpine plant cultivation

At the Royal Botanic Gardens, Kew, most of the bulbous plant collection and many alpine cushion-forming plants are grown in sand plunges. Many other alpine departments in other botanic gardens across the world also implement the same techniques. In these sand plunges, plants are buried in sand up to the pot rims, so their roots are essentially underground. Sand is normally used due to its sterile nature and because it is inexpensive and easy to purchase. The deeper the plunge area, the more effective it is at creating uniform humidity and temperature.

At home you could make a small sand plunge in a lined box (with drainage holes at the bottom) filled with builders' sharp sand. The process of adding plants is straightforward. The aim is to make as good contact as possible between the pot and sand. Therefore, you need to minimize and collapse any air gaps by firming the sand, and by watering the sand well after plunging the plant pots.

Your plants will benefit most from the properties of the plunge by being in porous terracotta pots. The plant roots will then have uninterrupted access through the terracotta to humidity from the sand. However, if you do not have terracotta pots, plants in plastic pots will still benefit from the uniformly cool root environment of being plunged around sand.

Plants gain the most from sand plunges during summer, when they can remain cool, and in winter, when the plunge can be watered to keep the fussier alpines slightly damp without having to water individual pots.

1. Choose a space for your pot (here of *Dionysia aretioides*) in the sand plunge and mark the size of hole required.
2. In the marked position, dig a hole in the sand to an appropriate depth, using a scoop or trowel. Take care not to spill the sand over neighbouring plants.
3. Insert the pot into the hole, making sure it is sitting on a solid base of sand.
4. Backfill around the pot edges, and firm in using the scoop.
5. Water the sand around the pot, to ensure it is well bedded in.

Lesser meadow rue

Thalictrum minus

Rocky places across Europe, Asia and North Africa, to 3,000m/10,000ft altitude, are home to this dainty wild flower. There, it forms a mound of deciduous pinnate foliage above which are airy clusters of pale flowers, 50cm/20in high.

—

WHERE TO GROW

Lesser meadow rue is perfect for a rock garden or raised bed. An open, partly sunny aspect is best in evenly moist but well-drained soil.

HOW TO GROW

Plant in spring in soil with added organic matter. Cut back in autumn, once the deciduous leaves have turned brown. Propagate by division in early spring (See Dividing an herbaceous perennial alpine, page 54).

GROWING TIP

Herbaceous plants such as lesser meadow rue are good companions for winter/early spring bulbs like *Crocus* that have the opposite growing season.

Family Ranunculaceae

Height & spread 25–30x30cm/ 10–12x12in

Flowering time Summer

Hardiness H6

Position Full sun–part shade

COW MEDICINE?
A 2016 study reported in the *Journal of Ethnopharmacology* found antimicrobial alkaloid chemicals in lesser meadow rue, supporting its traditional use as a herbal remedy for bovine mastitis.

Creeping thyme

Thymus praecox aka mother of thyme, wild thyme

Slowly creeping along the ground in Europe and Greenland is this miniature scented thyme. It forms a carpet of pink flowers, 10cm/4in high, attracting bees.

—

WHERE TO GROW

This mat-forming thyme species is perfect ground cover in open, sunny, evenly moist positions. It does not tolerate being overshadowed. Therefore, grow at the front edge of a rock garden or raised bed; it is also good for a pot or trough.

HOW TO GROW

When pot-grown, creeping thyme requires more watering than plants growing directly in the ground. Propagate by division in spring (see Dividing an herbaceous perennial alpine, page 54). Creeping thyme self-seeds quite freely after a few years.

GROWING TIP

Because creeping thyme is susceptible to crisping up in summer if not watered sufficiently, add organic matter when planting and keep the soil moist.

Family Lamiaceae

Height & spread
5–10x50cm/2–4x20in

Flowering time
Summer

Hardiness H6

Position Full sun

POLYMORPHIC ESSENTIAL OILS
There is polymorphism (meaning variation) in the essential oils produced by *T. praecox*. This varies geographically, so plants smell and taste different depending on where they are from.

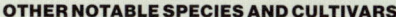

OTHER NOTABLE SPECIES AND CULTIVARS

- *T.* 'Doone Valley' is a variegated hybrid cultivar with pink flowers and zingy, lemon-scented foliage.
- *T. pulegioides* (broadleaved thyme) is another low-growing species, but with wider leaves dying back to the ground in winter.
- *T. serpyllum* (Breckland thyme) tolerates light foot traffic; it is closely related to *T. praecox*.

Candia tulip

Tulipa saxatilis

This long-lived tulip species is native to the rocky limestone slopes of Türkiye and the Greek southern Aegean islands. Garish bicoloured flowers, with pink petals and yellow centres, form on stems to 35cm/14in tall.

—

WHERE TO GROW

Grow in a well-drained rock garden or raised bed in full sun. Candia tulip is also suitable for a pot or trough. Ensure the plants are in an open position and not overshadowed by other plants when they are growing.

HOW TO GROW

In autumn, plant bulbs 10cm/4in below the soil surface, adding plenty of grit to the planting area. Water potted plants with liquid fertilizer every two weeks. Lift and separate bulb clumps in autumn (see Division, page 40), to spread to other positions in the garden. Candia tulip is susceptible to viruses (see Pests and diseases, page 135).

GROWING TIP

Watch out for squirrels and mice digging up and eating the bulbs; protect planting areas with chicken wire if this is likely to be a problem.

Family	Liliaceae
Height & spread	20–30x10–40cm/ 8–12x4–16in
Flowering time	Spring
Hardiness	H6
Position	Full sun

A TULIP FOR LIFE
Unlike many of our favourite tulip cultivars that diminish after a few years in the ground, the Candia tulip, when planted in the right position, will keep increasing and spreading over many years, to form a large colony (see page 21).

Violet

Viola

A common occurrence in the mountainous landscape are groups of these colourful little annuals and herbaceous perennials, which often have heart-shaped leaves and very bright, bilateral, scented blooms in spring and summer.

—

WHERE TO GROW

Many species are suitable for cultivation in moist but well-drained soil in full sun or part shade in a rock garden or raised bed. More delicate species (less tolerant of damp) are also good in a pot in an alpine-house sand plunge (see Using a sand plunge for alpine plant cultivation, page 128).

HOW TO GROW

Plant in soil that has been improved with a handful each of grit and garden compost, to increase drainage and moisture retention. Propagate perennial species by division (see Dividing an herbaceous perennial alpine, page 54) and annual ones by seed sown in autumn–spring (see Growing alpines from seed, page 60).

GROWING TIP

Some of the species described below (*V. biflora,* *V. calcarata*) are true alpines and need cool conditions to thrive, so protect them from hot summer sun by planting near rocks in part shade.

Family Violaceae

Height & spread
2–10x5–30cm/
¾–4x2–12in

Flowering time
Spring–summer

Hardiness H5–H7

Position Full sun–part shade

ROSULATE VIOLAS
In the high-altitude regions of Patagonia, *Viola* has developed into highly specialized 'rosulate' forms, where the leaves are arranged in tight concentric circles forming a rosette. Nothing but the distinctive *Viola* flowers give them away.

Viola odorata

Viola calcarata

NOTABLE SPECIES

- *V. biflora* (twin-flowered violet) is a small, yellow-flowering perennial from the European Alps and only 10cm/4in tall.
- *V. calcarata* (mountain violet) is clump-forming perennial from the calcareous soils of the European mountains, with vivid purple flowers in spring; grows best in cool conditions.
- *V. jooi* (Carpathian violet), from the Carpathian Mountains of Romania and Ukraine, is a clump-forming species with runners and bears soft lavender flowers in summer.
- *V. odorata* (sweet violet) is not a true alpine species; however, this vigorous perennial is a good starting option for a rock garden or raised bed. It has scented, dark purple blooms in spring.

Troubleshooting

HOLES IN THE MIDDLE OF PLANTS

As a part of the aging process, alpine plants often develop gaps and holes in their growth. This is perfectly natural, and common in herbaceous perennials, cushion- and mat-forming species. You can lessen the chance of this occurring by ensuring plants always have access to good light and are never covered over by fallen leaves or other plants. If you find the holes unsightly, it is the time to consider propagating the plant (see page 34) and/or replacing it.

ROTTING PLANTS

Plants with thickened taproots or dense top growth may be particularly prone to rotting. Always test the soil for moisture levels with your finger before watering blindly, and water with care in cool weather. Water plants from the edge, not overhead, to keep their foliage dry. Ensure containers are raised up off the ground on pot feet or bricks so they drain after watering. See also Watering, page 31.

PESTS AND DISEASES

Because of the slow-growing nature of alpine species, many do not suffer from extreme pest problems in the same way that plants with faster growth rates do. They can, however, be affected by many of the same types of pest that you find in the rest of the garden and under glass. These are a natural part of the ecosystem, and many pests don't cause lasting damage and are easily deterred by growing strong, healthy plants.

One of the most damaging pests for alpines can be the mollusc. Watch out for slug and snail slime trails and half-eaten flowers, tubers and shoots. Protect from damage on spring flowers and foliage by regularly inspecting not only outdoor areas with a torch in the evening, but also checking under and around rims of pots and containers. You can also set beer traps or leave upturned pots in the evenings to collect molluscs to dispose of in the morning. Another solution is to apply biological control in the form of nematodes (particularly the species *Phasmarhabditis hermaphrodita*), which are effective at killing slugs when they hide under the soil. The nematodes are active only in the temperature range 5–20°C/41–68°F, and they require a thin film of water to survive, so must be applied mixed into a water solution in spring or autumn. This is effective in containers and in the ground.

Another general pest to combat is the tortrix moth caterpillar. Watch out for the tell-tale, nibbled leaf edges and curled leaves bound by strands of silk, which hide this caterpillar and its cocoons. Manually inspect plants regularly and remove caterpillars and cocoons.

Sempervivum leaf miner causes damage to the outer leaves of houseleeks. Unlike rotting, it may affect only a few plants in a pot. The larvae eat the leaves from the inside, causing the hollow appearance seen here.

The symptoms of viruses in bulbous plants include irregular streaking and distortion of foliage, here on snowdrops (*Galanthus*).

Birds and mammals can disturb soil by digging. If this is a repeated problem, consider temporarily covering your alpines with netting or installing some inexpensive, motion-activated animal deterrent devices.

Many monocot species (mostly bulbs) are susceptible to viruses, which are often spread by sap-sucking insects such as aphids. Watch out for contorted or irregularly streaky growth. Immediately dispose of any plants you think might have viruses, as they are not treatable. Make sure you sterilize cutting blades regularly, to prevent cross-contamination of plants with viruses when pruning or cutting.

Plant-specific pests
Narcissus bulb fly can be an issue for bulbs in the family Amaryllidaceae including daffodils (*Narcissus*) and snowdrops (*Galanthus*). The adult flies lay eggs at the bases of leaves, and the larvae can eat the bulbs from inside. Lift and inspect bulbs if your plants stop regularly flowering and their leaves look weak and stunted. Infected bulbs will have a hole at the base and be squishy with a maggot inside. Dispose of infected bulbs.

Sempervivum leaf miner, as the name suggests, infects only houseleeks (*Sempervivum*). This is a species of hoverfly

(*Cheilosia caerulescens*) that lays its eggs on the leaves of houseleeks. The eggs hatch and the larvae eat their way into the leaves from the outside of the rosette. The damage looks very similar to rotting overwatered plants (see page 134) but, instead, you should find the larva inside the leaf. Remove all infected leaves from the plants and consider discarding whole rosettes or the contents of pots if the infection is bad. This pest is native to continental Europe and is uncommon in the UK and USA.

Pests under glass and in pots
Plants growing in containers, especially in greenhouses, are subject to a higher risk of pest infestation. This is because they are in conditions that are warmer than outdoor conditions and they are protected from wind. Thus they produce soft edible green growth, which pests such as aphids and mealy bugs can't resist. Conversely, plants in containers can also be stressed by underwatering, making them vulnerable to pests.

There are a few easy ways to encourage strong plant health in containers and under glass: avoid overwatering (see page 31) and excessive feeding (see page 31); and make sure growing spaces are well ventilated and plants are well spaced, so they have access to good light levels and the greenhouse glass is clean. You can purchase automated, vent-opening systems for greenhouses.

An aphid or red spider mite outbreak may occasionally occur, particularly in very sheltered conditions, despite good cultural conditions. Watch out for aphids particularly in the centre of cushion-forming plants, where they can hide undetected until the plant shows signs of wilt. You can manually squash aphids or wash them off with jets of water. Move affected plants outside into the wind and rain; this will often negatively impact the growth rate of the aphids or red spider mites. As a last resort, dilute biodegradable dish soap solution can be effective at killing red spider mites and aphids under glass, by blocking their breathing pores. Mix one drop per 1 litre/1¾ pints and spray on to every surface of the foliage. This solution can damage foliage in full sun or if applied more than once per month. Only use this inside a greenhouse, and never in the garden/near water as it can negatively affect other insect and aquatic life.

Vine weevil is a pest that can detrimentally affect container-grown alpine plant collections. When the weather is warm in spring and summer, watch out for adults: they are 1–2cm/½–¾in long, six-legged, black beetles with ridged backs and yellowish markings. The damage on foliage is very easily identifiable as geometric notches on the edges of plant leaves. You can buy traps for the adults or collect them up at night.

Vine weevil larvae can become active in pots in early spring and autumn. Keep an eye on potted plants of species that have fleshy roots or are particularly susceptible, such as saxifrage (*Saxifraga*), houseleek (*Sempervivum*), stonecrop (*Sedum*) and *Primula*. Nematode species such as *Steinernema kraussei* or *Heterorhabditis megidis* are very effective at killing the larvae but need temperatures above 5°C/41°F to be effective. Nematodes are

The small, pale yellow vine weevil larvae cause damage by eating roots and underground stems, here on houseleek (*Sempervivum*).

Root mealy bugs have a white woolly appearance, and they suck on the roots of potted plants.

a much better treatment option than pesticides as they are harmless to the environment, pets and people. In the middle of winter, you may need to consider bare-rooting plants and replacing soil, to remove larvae. As a very last resort, you can apply a soil-drench insecticide (such as a systemic neonicotinoid) to the potted pots.

Root aphids and *root mealy bugs* may also attack containerized plants, especially those under stress. These subterranean pests don't usually kill plants unless they are already weak. They sit on the roots, usually at the edge of a pot and

suck sap. Check potted plants from time to time by tipping them upside down and examining the roots. Because root aphids and root mealy bugs are relatively hard to treat, the best course of action is to try propagating the plant from above-ground cuttings or to transfer the affected plant into the garden. Root mealy bugs and root aphids do not survive as well without the protected pot environment.

What to do when

AUTUMN

Although growth is winding down in this season, some alpine species take the opportunity to flower before the winter sets in. This is particularly true of Mediterranean bulbs. Species such as *Crocus, Cyclamen* and *Colchicum* will be flowering between late summer and late autumn.

- Plant early-flowering bulb species such as snowdrop (*Galanthus*), *Narcissus*, grape hyacinth (*Muscari*) and some squill (*Scilla*).
- Collect seeds from summer-flowering species.
- Consider lifting and dividing some of the herbaceous perennial species (see Dividing an herbaceous perennial alpine, page 54).
- Clear up autumn leaves as they fall.
- Cut back brown herbaceous perennials.
- Ensure your bulbous plants are not smothered in pests as they emerge (see page 134). Pests can overwinter in foliage and excessive leaves from nearby deciduous trees can cover delicate alpines.
- Consider winter protection for some alpine species. This could involve lifting and potting up more tender plants such as some aloes and bringing pots under glass into a greenhouse. It could also mean erecting a rain shelter for delicate species planted in the ground or in a trough that is too heavy to move (see Winter protection, alpine houses and sand plunges, page 22).
- Sow seeds, especially species that require a cold period to break dormancy (see Growing alpines from seed, page 60).

WINTER

The majority of alpine species will now be dormant, although there will still be some flowering plants: for example, snowdrops (*Galanthus*), some squill (*Scilla*) and some *Crocus*.

- Keep an eye on any winter-protection structures and greenhouses (see Winter protection, alpine houses and sand plunges, page 22); regularly check for leaks to ensure your plants are dry.
- Towards the end of winter, consider investing in a water butt to collect and store rainwater over the coming season. Rainwater is particularly appreciated by plants in pots as it does not contain the minerals and chlorine that tap water does, particularly if you live in a hard-water (calcium-rich) area.
- As long as the ground is not frozen, decide if there are any herbaceous perennials you would like to lift and divide (see Dividing an herbaceous perennial alpine, page 54).

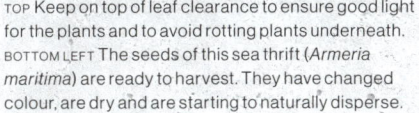

TOP Keep on top of leaf clearance to ensure good light for the plants and to avoid rotting plants underneath. BOTTOM LEFT The seeds of this sea thrift (*Armeria maritima*) are ready to harvest. They have changed colour, are dry and are starting to naturally disperse.

ABOVE When cutting back in autumn, ensure you use clean secateurs and remove dead flowering stems at the bases.

Some frost-sensitive plants, such as this lace aloe (*Aristaloe aristata*) and agaves, may require the additional winter protection of a cold frame.

- In late winter, plant early flowering alpine species, and repot any pot-bound containerized specimens (see page 32).

SPRING

Spring is the most vibrant season of alpine gardening.

- Make sure you look at your plants daily so you don't miss a single bloom. Many species bear individual flowers that may last only a few days.
- Wait until the risk of frost has passed and the worst of the relentless heavy rain has subsided before removing winter protection.
- Feed your container-grown plants every few weeks with a liquid fertilizer or apply a granular fertilizer to the soil (see Feeding, page 31).
- Continue to monitor soil moistness of container-grown plants, which can dry quickly on sunny days in early and mid-spring (see Watering, page 31). Take care not to overwater – some plants can flower greatly in early spring, while it is still cold, before their demand for water increases later in spring.
- Repot pot-grown plants, watering carefully until they are vigorously growing (see Repotting, page 32).
- Deadhead regularly to encourage more blooms.
- Prevent the growth of mould, particularly on plants with tight growth such as cushion-forming

plants, by ensuring good air circulation as well as by deadheading.

- Pests can boom in late spring as the weather warms. Watch out for aphids on soft new growth; these can be problematic when plants are grown under glass (see Pests under glass and in pots, page 136).

SUMMER

The main cultural worries in summer are high temperatures and drought.

- Protect a greenhouse with shade netting or shade paint on the glass, and move troughs to a position where they are not in full sun all day.
- Monitor the growth of plants; some alpine species will experience a slowing of growth and even dormancy in the hottest part of the year, so may not take up much water. Test 5cm/2in down into the soil and water only when it feels dry to the touch. This could be as often as every few days or as little as once every couple of weeks.
- Keep cutting small bunches of flowers to bring into the house (see Making an alpine cut-flower display, page 76).
- Monitor how many flowers you have over the growing season. If there are gaps in early and midsummer in flowering alpine species, consider adding some Mediterranean species that bloom throughout summer, to provide colour and interest. Plant these in autumn or the following spring.

Alpine troughs will needed to be watered frequently during dry spells in the summer.

Index

Quarto

First published in 2024 by Frances Lincoln,
an imprint of Quarto.
One Triptych Place,
London, SE1 9SH
United Kingdom
T (0)20 7700 6700 F (0)20 7700 8066
www.Quarto.com

A catalogue record for this book is available from the
British Library.

ISBN 978-0-7112-9044-0
eISBN 978-0-7112-9045-7

10 9 8 7 6 5 4 3 2 1

Typeset in Adobe Garamond and
Neue Haas Grotesk Display
Design by Arianna Osti

Printed in China

Author's acknowledgements

I would like to thank Olivia and my mum and dad as well as my whole family for endless encouragement, not just in this book, but with everything. You have tirelessly put up with me and have always helped behind the scenes with photos and tasks. Thank you to Viv for your support and proofreading. I would also like to thank the team at RBG Kew for encouragement and allowing me access to the alpine potting shed and helping me take photos. Thank you, Tom Freeth, for your stunning photos and always being supportive. Thank you Raz Chisu, Chris Grey-Wilson and Tom for instilling your knowledge and passion for alpines on our travels across Europe; many photos from these adventures feature in the book. Also, to Tony Bryan and the Alpine Garden Society for encouragement and supporting me to be a trustee of the society.

Joanna Chisholm has done an amazing job editing my ramblings, Arianna Osti has made the book look beautiful, Pei Chu has digitised stunning botanical illustrations and Michael Brunström and Lydia White have been very encouraging throughout the whole project, thank you all.

Instagram: @matthewjjeffery

Photographic acknowledgements

All photos Matthew Jeffery except for:
Tom Freeth 2, 21, 23 bottom, 96
Shutterstock: 79, 90, 125, 130